Quantifying the Impact of Technical Barriers to Trade

SIE *Studies in International Economics* includes works dealing with the theory, empirical analysis, and evaluation of international policies and institutions, in the area of international macroeconomics and finance, international trade theory and policy, as well as international legal and political economy.

Keith E. Maskus, Peter M. Hooper, Edward E. Leamer, and
J. David Richardson, Editors
Quiet Pioneering: Robert M. Stern and His International Economic Legacy

Bjarne S. Jensen and Kar-yiu Wong, Editors
Dynamics, Economic Growth, and International Trade

Jagdish Bhagwati and Mathias Hirsch, Editors
The Uruguay Round and Beyond: Essays in Honor of Arthur Dunkel

Kala Marathe Krishna and Ling Hui Tan
Rags and Riches: Implementing Apparel Quotas under the Multi-Fibre Arrangement

Alan V. Deardorff and Robert M. Stern *Measurement of Nontariff Barriers*

Thomas Cottier and Petros C. Mavroidis, Editors
The World Trade Forum, Volume I: State Trading in the Twenty-First Century

Rajesh Chadha, Sanjib Pohit, Alan V. Deardorff, and Robert M. Stern
The Impact of Trade and Domestic Policy Reforms in India: A CGE Modeling Approach

Alan V. Deardorff and Robert M. Stern, Editors
Constituent Interests and U.S. Trade Policies

Gary R. Saxonhouse and T. N. Srinivasan, Editors
Development, Duality, and the International Economic Regime: Essays in Honor of Gustav Ranis

Charles P. Kindleberger *Essays in History: Financial, Economic, Personal*

Keith Acheson and Christopher Maule
Much Ado about Culture: North American Trade Disputes

Alan V. Deardorff and Robert M. Stern *Social Dimensions of U.S. Trade Policies*

Thomas Cottier and Petros C. Mavroidis, Editors
Regulatory Barriers and the Principle of Non-Discrimination in World Trade Law

Bernard Hoekman and Jamel Zarrouk *Catching Up with the Competition: Trade Opportunities and Challenges for Arab Countries*

Jun Fu *Institutions and Investments: Foreign Direct Investment in China during an Era of Reforms*

Robert M. Stern, Editor *Services in the International Economy*

Keith E. Maskus and John S. Wilson, Editors
Quantifying the Impact of Technical Barriers to Trade: Can It Be Done?

Quantifying the Impact of Technical Barriers to Trade: Can It Be Done?

Keith E. Maskus and John S. Wilson,
Editors

Ann Arbor

THE UNIVERSITY OF MICHIGAN PRESS

Published in the United States of America by
The University of Michigan Press
Manufactured in the United States of America
∞ Printed on acid-free paper

2004 2003 2002 2001 4 3 2 1

A CIP catalog record for this book is available from the British Library.

Library of Congress Cataloging-in-Publication Data applied for
ISBN 0-472-11247-3 (cloth : alk. paper)

Contents

Preface

Tension in the international trading system is increasingly focused on standards and technical regulations. Developed and developing countries have expressed concern that their exporters find it ever more difficult to comply with regulations and safety standards set by importing nations. Technical barriers to trade may be, in fact, attempts to restrict market access through rules imposed on product characteristics and production processes. Authorities in importing nations, however, defend such practices. They often argue that deep regulation is necessary to ensure food safety, the health of animal herds and crops, consumer welfare, and integrity of telecommunications networks. The trade conflicts that such differences engender are growing in frequency and importance.

It is remarkable how little is known about the trade impact of technical barriers based on empirical data and analysis. Economists cannot say with confidence whether such restrictions tend to reduce trade by virtue of raising compliance costs or expand trade by increasing consumer confidence in the safety and quality of imported goods. Claims abound on both sides, about which little illumination is available from existing studies.

Despite the lack of systematic empirical data on trade and standards, important policy decisions are needed about what constitutes acceptable norms and rules in this area. For example, multilateral negotiations at the World Trade Organization have set out rights and obligations for members regarding sanitary and phytosanitary measures that countries use to protect human, animal, and plant health. Authorities in regional trading arrangements, such the Free Trade Area of the Americas (FTAA) and the Asia Pacific Economic Cooperation Forum (APEC) are discussing ways to remove technical and regulatory barriers. Some countries engage in bilateral cooperative mechanisms to reduce compliance costs, which could have the effect of excluding products from third countries. Again, there is virtually no systematic information about the potential costs and benefits of such regimes.

The motivation for this volume, therefore, centers on providing the reader with a survey of current and anticipated approaches to quantifying the trade impact of technical regulations. We are particularly interested in empirical research and quantification methods that are practical and provide concrete links to public policy decisions. Measuring the trade effect of standards is a complicated task. There are an infinitely varied array of regulations and channels through which these might affect trade.

For example, safety rules are often product-specific and rely on the engineering and health expertise of professionals in different countries. In this context, an exporter faced with satisfying the requirements of varying rules in different foreign markets faces a complex set of choices. Should the exporter develop a single design that may be modified for particular markets, or a unique design for each individual market? To what extent does this choice affect decisions on where to export? Are the costs of meeting foreign regulations essentially one-time, fixed design costs or on-going, variable factor and compliance costs? Numerous other questions could be posed in this context.

The World Bank organized a conference in April, 2000 in which leading trade specialists were invited to provide their perspective on how to quantify the effects of technical regulations on international trade. This volume brings together contributions from participants in the conference. The research and policy discussions presented here constitute a comprehensive and state-of-the-art review of issues relevant the economic analysis of trade-related technical barriers. The papers range from discursive policy discussions to empirical analyses of particular standards. They cover such aspects as how the WTO and other organizations address technical regulations, to the economic benefits of standards, and methods for considering whether regulations are legitimate policy tools or protectionist devices.

Several of the chapters set out conceptual and technical frameworks for considering the effects of standards on costs, consumer demand, and trade, both in theoretical and applied contexts. One chapter considers how bans by the European Union on production and import of genetically modified foods might influence trade costs and trade volumes, finding that the impact on agricultural exporters could be severe. Another chapter points out that the trend among industrialized nations to recognize each other's testing procedures, but not those of developing countries, poses risks for the emergence of a bifurcated international trading system. Still another points to numerous practical difficulties in measuring costs of compliance with environmental regulations. All the contributions provide practical advice on methods for determining such costs and estimating their trade impacts.

Acknowledgments

It is impossible to put together such a large project without the assistance of numerous people, to whom we wish to extend our sincere gratitude. A special thanks is extended to the World Bank's trade team and assistants. Josef M. Ritzen of the World Bank chaired the conference at which many of the papers in this volume were presented. Tsunehiro Otsuki of the World Bank provided outstanding professional assistance in preparation for the conference and with papers presented in this volume. Lisa Johnson assisted with conference organization and Robb Simms of the Bank prepared the manuscript for publication. Tony Sihsobhon and Polly Means of our IT Support Group at the Bank assisted us with computing resources. Microsoft Corporation provided us with Publisher 2000 Deluxe® software used to produce this volume. The Bank's distance learning team and videoconference staff merit a special note of gratitude. Robin Koenigsberg and Eina Wong at the University of Colorado provided timely research assistance in the preparation of manuscripts. We are particularly indebted to the Department for International Development (DFID), United Kingdom, for its generous financial support.

Food Policy for the tables, "Composition of Processed Food Exports from Developing Countries", and "Comparative Growth Rates of Processed Food and Primary Commodity Exports, 1970-1994," from the article, *Processed Food Exports from Developing Countries: Patterns and Determinants* by P.C. Athukorala and K. Sen. ©1998 with permission from Elsevier Science.

Brookings Institution Press for the Richard E. Baldwin article, "Regulatory Protectionism, Developing Nations and a Two-Tier World Trade System" from *Brookings Trade Forum: 2000.* ©2001 The Brookings Institution.

Disclaimer

All opinions expressed by the authors are strictly personal and should not be attributed to any government, official agency, or institution with which they may be, or have been, affiliated.

Keith E. Maskus and John S. Wilson

Tables and Figures

Tables

Figures

Abbreviations and Acronyms

APEC	Asia Pacific Economic Cooperation
APHIS	Animal and Plant Health Inspection Service (US)
ARSO	African Regional Standards Organization
BSE	Bovine Spongiform Encephalopathy
CAP	collective action plan
CEN	Comité Européen de Normalisation
CENELEC	Comité Européen de Normalisation Electrotechnique
CGE	computable general equilibrium (models)
COMESA	Common Market for Eastern and Southern Africa
DSB	Dispute Settlement Body (WTO)
EC	European Community
EEA	European Economic Area
EFTA	European Free Trade Area
ETSI	European Telecommunications Standards Institute
EU	European Union
FAO	Food and Agriculture Organization
FDA	Food And Drug Administration (US)
FDI	foreign direct investment
FMD	Foot-to-Mouth disease
GATT	General Agreement on Tariffs and Trade
GDP	gross domestic product
GM	genetically modified
GMOs	genetically modified organisms
GNP	gross national product
GSP	generalized system of preferences
GTAP	Global Trade Analysis Project
IADB	Inter-American Development Bank
IAP	individual action plan
IEC	International Electrotechnical Commission
IRTS	increasing returns to scale
ISO	International Organization for Standardization
IT	information technology
ITU	International Telecommunications Union
JAS-ANZ	Joint Accreditation System – Australia and New Zealand
MENA	Middle East and North Africa
MFN	most favored nation
MRA	mutual recognition agreement(s)

MVA	manufacturing value added
NAFTA	North American Free Trade Agreement
NGOs	non-governmental organizations
NSB	national standards bodies
NTB	non-tariff barriers
OECD	Organization for Economic Co-operation and Development
PTA	preferential trade arrangement
RCA	revealed comparative advantage
SABS	South African Bureau of Standards
SDO	standards development organizations
SIC	Standard Industrial Classification
SITC	Standard International Trade Classification
SPS	sanitary and phytosanitary standards
SSA	Sub-Saharan Africa
TABD	transatlantic business dialogue
TBT	technical barriers to trade
TRAINS	Trade Analysis and Information System
UNCED	United Nations Conference on Environment and Development
UNCTAD	United Nations Conference on Trade and Development
UNIDO	United Nations Industrial Development Organization
USDA	United States Department of Agriculture
VER	voluntary export restraint
WTO	World Trade Organization

Note: Not all abbreviations used in this volume are listed above.

CHAPTER 1

A Review of Past Attempts and the New Policy Context

Keith E. Maskus
John S. Wilson

Introduction

The impact of standards and technical regulations on trade is at the forefront of global policy discussions.[1] Prominent recent examples of trade disputes involving technical regulations among the industrialized nations include Japanese sanitary requirements for imported varietal fruits, the European Union's ban on imports of beef raised with growth hormones, and the French refusal to import asbestos-containing products. Looming in the near future is the issue of trade restraints on genetically modified foods, which promises to be particularly contentious and involve nations at all levels of development.

The potential trade-inhibiting effects of technical regulations poses a particular challenge to developing nations as they seek to increase production for export markets. In regard to voluntary standards, such as those in the International Organization for Standardization (ISO) 9000 series on quality, developing nations face constraints in absorbing best-practice information on standards and mobilizing resources necessary to adopt appropriate process and production methods (World Bank 2000b). In addition, as traditional trade protection measures such as tariffs, quotas, and voluntary export restraint (VER) agreements have been liberalized, barriers to trade reflected in domestic technical regulations have become more important channels through which trade is blocked.[2]

The successful conclusion of the Uruguay Round in 1994 highlighted the importance of national standards as non-tariff barriers (NTBs). Not all market-access commitments in the World Trade Organization (WTO) Agreements have been fully implemented or enforced.[3] Neither have other barriers to trade, such as subsidies to agricultural production or prohibition on foreign investment in services, been completely eliminated. It is clear, however, that domestic regulations affecting imports through technical requirements, testing, certification, and labeling represent one of the most important new areas of focus in continuing liberalization efforts. This chapter outlines why this is the case and also discusses the need for quantitative analysis in the area.

Mandatory regulations imposed by governments at the border can produce serious distortions in commercial markets. For example, domestic regulatory systems may restrain trade and limit market entry through environmental, health, or safety mandates not based on international norms. These requirements may also be discriminatory within the context of WTO disciplines, including commitments undertaken by WTO members in the Agreement on Technical Barriers to Trade (TBT) or the Agreement on the Application of Sanitary and Phytosanitary Standards (SPS).

Technical regulations imposed on traded goods and agricultural products affect trade patterns, the ability of producers to enter new export markets, and consumer costs. While these claims are valid in principle, there is relatively little empirical evidence about technical regulations and trade. Such regulations include specific performance requirements, such as how much pressure brake pads in automobiles can withstand before they must be replaced. They also involve conformity assessment rules, including testing, certification, laboratory accreditation, and related labeling requirements for products.[4] Few rigorous empirical attempts have been made to determine the economic impact of differing regulations and conformity assessment requirements. Important questions remain unanswered and evidence about them could directly affect trade policy choices and the success or failure of liberalization efforts in which developing countries have a direct stake.

Adding to this context, requests by private industry to focus on costs of regulatory barriers to trade have accelerated over the past decade. For example, the Transatlantic Business Dialogue (TADB) and much of the work in the Asia Pacific Economic Cooperation (APEC) talks has been focused on finding ways to reduce technical barriers.[5]

The World Bank is beginning a multi-year, multi-country study of these questions. As a first harvest of this work, the Bank held a workshop on April 27, 2000 in which several economists presented current research on aspects of standards, regulations, and trade. Authors were also asked to draw conclusions about the potential for empirical research in the area and to note particular difficulties in this area. This volume contains those papers, which we overview in a later section. They are the product of that thinking and reflect the current state of the art in theoretical and applied work.

In this introductory chapter we start with an analytical overview of the policy debate and methodological issues surrounding product standards and technical barriers to trade. This is presented as a step toward providing a framework for new empirical work on standards, the subject of the second chapter. In the next section we discuss essential motivations for performing research in this area. While information to be gathered in any study of these issues should be of considerable value in itself, the ultimate objective is to make that information useful for national policymakers and international trade negotiators.

In the third section we examine the policy context driving the need for new data and empirical evidence. We discuss the rising incidence of technical regulations as instruments of commercial policy in trade disputes. This issue is important for enterprises in both developed and developing nations. However, we emphasize the particular concerns of developing countries in the debate over such regulations. Firms in developing countries must shoulder significant costs in meeting mandatory government standards to serve export markets. In addition, developing countries have in place domestic regulatory policies affecting imports that continue to act as protectionist barriers shielding inefficient domestic producers from competition.

In the fourth section we review the role of standards in international trade and discuss policy justifications for reaching voluntary standards and imposing mandatory technical regulations. These justifications rest on the need to address such market failures as asymmetric information about product quality, inadequate provisions to safeguard public health, and the risk of network incoherence. We note particular issues that arise in the cross-border context. Standards have impacts on both static and dynamic market failures, suggesting there are subtle efficiency and welfare tradeoffs.

In the fifth section we introduce the papers in this volume, reviewing their objectives and key results. Taken together, they provide a comprehensive statement about current analytical issues regarding technical regulations, standards, and trade. We use their insights to inform empirical approaches, set out in the next chapter. Our ultimate purpose is to identify techniques that are practical and may be fruitfully extended to the empirical analysis of standards and trade. Such steps would include administering firm-level surveys in developing countries and establishing econometric and simulation approaches that could be applied to micro data for understanding the role of regulations in influencing exports. We provide brief concluding remarks in the final section.

Reasons to Quantify the Effects of Standards and Regulations

As we discuss in the following section, there is considerable interest within the global policy community in solid empirical work on the impacts of technical regulations affecting trade. In the main, this interest reflects the belief that regulations often constitute important non-tariff barriers to trade and that their use is proliferating. Many observers claim that the trade-restricting effects of technical regulations in the developed countries are particularly costly for exporters in developing nations. These arguments arise from anecdotal examples and case studies. However, there is little systematic evidence about them. Thus, a primary motivation for study of these issues is simply to shed light on such claims. How pervasive are technical regulations that affect trade? How do exporters and importers alter their cost structures and production lines to accommodate such requirements? Are technical barriers a reason that firms choose not to enter export markets? Are inspection procedures and labeling requirements discriminatory? Evidence on these and related questions, gathered from detailed firm-level surveys, would be of great interest to firm managers, researchers, and policymakers. For example, firm managers could use the catalogue of information developed to assess their own strategies in dealing with standards and regulations.

Ultimately, however, what matters are the policy uses to which the information will be put. We envision the techniques and results of the analysis to become inputs into programs to make more rational the complex tapestry of domestic and international regulations and standards. The key concept is "rationalization" in preference to "elimination". The former approach may be designed to facilitate trade without sacrificing important social and economic objectives. The latter approach, in some cases, may be economically inappropriate and socially costly.

To see this, consider that standards and regulations differ fundamentally from taxes and quotas on trade. Such classic trade barriers are inefficient and discriminatory taxes on foreign sources of economic activity. As such, they raise costs to consumers and input users, inefficiently allocate resources, and protect entrenched domestic market power. Economic analysis demonstrates that countries mutually benefit from negotiating their removal or reduction. For their part, regulatory barriers exist in principle to achieve important objectives that would go under-served in the private market, such as public-health maintenance or environmental protection. Elimination of such regulations could produce social losses in excess of any economic efficiency gains.

In fact, there is more subtlety to this distinction. Tariffs and quotas may well exist for purposes of economic and social regulation. However, because they are indirect means and embody a protectionist and discriminatory element, they erect costs that would not arise from non-discriminatory regulations aimed directly at the underlying goals. Thus, it has been relatively straightforward for WTO members to establish the principle that border restraints are inappropriate methods of regulation and should be removed if there are more direct means available for achieving regulatory purposes. [6]

In principle, standards and regulations directly aim at overcoming market failures, as we discuss in Section 4. Thus, unless they discriminate between sources of supply, they do not embody secondary trade costs and may be the least-trade-restricting policies available. Indeed, under some circumstances, standards could expand trade. Any efficiency costs of standards are the investment societies make in achieving beneficial regulation. In consequence, it is not clear that the trade impacts of technical barriers are inefficient or that they should be the subject of multilateral negotiations. Put another way, removing technical regulations on trade would not necessarily achieve efficiency gains sufficient to overcome losses from weaker social protection. There is no standards analogue to the claim that free trade in goods is globally optimal.

Despite this benign view of technical regulations in the abstract, history demonstrates that countries may use them for questionable purposes in practice. Thus, regulations may discriminate against foreign suppliers, both in their construction and in their outcomes. They may be used to gain strategic international trade advantages for domestic firms over foreign competitors. They may be non-transparent and needlessly force firms to duplicate certification costs. They may be written to exclude both domestic and foreign entrants into a particular market, serving to support entrenched monopolies. Finally, they may be stronger than necessary to achieve a particular level of social protection, imposing excess costs on consumers and using industries.

Thus, rationalizing costly regulations can be beneficial. Such rationalization would entail ending discriminatory treatment, removing duplicative testing requirements, recognizing that foreign standards could achieve the same level of social or consumer protection as domestic standards, making regulation more transparent, and scaling them at levels that do not impose excess costs on consumers and firms.

In this context, the results of serious and extensive empirical inquiry could facilitate work on a number of important trade questions, beyond the fundamental one of discovering how pervasive and costly regulations are in practice. First, quantification of trade effects and production costs should provide an initial step toward supporting calculation of the welfare costs and efficiency impacts of standards and regulations. Such computations should determine whether regulations are more or less costly than classic border restraints. Second, the results should help inform governments as to whether their regulations depart significantly from the least-trade-restricting standards available for a given policy objective. In turn, governments could be urged to adopt less restrictive policies. Third, the information should provide a framework within which governments can assess the damages to their country's trade benefits from foreign regulations that may violate associated WTO obligations. In turn, they could serve as a basis for calculating compensation claims and for buttressing arguments in panel disputes. Finally, more information on how regulations operate could help resolve and limit future trade disputes. These ideas are developed further in the next section.

Standards and the New Trade Agenda: Why Economic Analysis Matters

There are a number of reasons why empirical work on technical and regulatory barriers to trade is particularly important in relation to the new trade agenda. Demand for such analysis and data is driven in part by significant curtailment of tariffs and traditional NTBs. We know considerably more about how tariffs and protective barriers such as quotas affect trade than we do about how technical barriers influence trade. For example, it is estimated that reducing tariffs and export subsidies by 40 percent in agriculture would produce a $60 billion per year rise in real global income (Hertel, et al 1999). Moreover, Hertel and Martin (1999) found that reducing applied tariffs on manufacturers by 40 percent would increase the volume of world trade by $380 billion. These data help inform policy tradeoffs and negotiating priorities. There are demonstrable benefits for developing nations from liberalizing barriers to goods trade, including NTBs related to regulation at the border. Developing countries account for 20 percent of world total exports in goods, with manufacturing goods totaling approximately 70 percent of their overall exports. Reforming duplicative or discriminatory technical barriers, either through changes in the mandatory standards themselves or through alterations in testing and certification requirements, should increase trade and economic efficiency.

The importance of empirical work on standards is also reflected in the rising incidence of technical regulations as instruments of commercial policy. Examples of disputes brought to the WTO and other institutions highlight this fact. It is evidenced further by increased calls for harmonization of standards and negotiation of mutual recognition agreements (MRAs) covering product testing and certification systems. The industrialized nations have dominated international policy deliberations on standards and account for the bulk of MRAs. However, in both of these areas developing countries have a special stake, as they have had little involvement in the development of international standards and lack much of the infrastructure necessary to support the modern testing procedures that mutual recognition of testing results requires.

The Multilateral Trade System: Standards as a Priority for Debate

There was a marked rise in the number of trade disputes over standards brought to the WTO during the past five years. The majority of these cases centered on trade in agricultural products and obligations under the SPS

Agreement. The highest-profile standards cases involved agricultural production and trade among the developed countries, such as the dispute between the European Union and United States over hormone treated beef.[7] The use of genetically modified organisms (GMOs) in agriculture also has raised tensions that must be addressed. Among the central questions in this debate are what constitutes an "international" standard and what is "least trade distorting" within the context of WTO disciplines.

Claims that countries have violated provisions of the TBT Agreement have also increased since 1995. Disputes over domestic regulations affecting process and production methods through which goods are manufactured are becoming more frequent. The United States and European Union, for example, have been debating EU draft regulations on waste from electronic and electrical equipment. These regulations would affect the use of chemicals and other inputs employed to produce a wide range of consumer electrical products. As a consequence, the viability of WTO disciplines to address areas where domestic regulation affects trade is being tested. This situation will undoubtedly become more pronounced as economic activity disperses further around the world and exports from developing countries of finished goods continue to grow. These countries already have status to bring complaints in the WTO, as will new entrants over the next few years such as China, Russia, and Ukraine.

As of October 2000, the WTO Dispute Settlement Body had considered or received a total of 27 disputes that referenced either the SPS or TBT Agreements.[8] These cases are listed in Table 1.1. In procedural terms these cases ranged from requests for consultations through panel and appellate body rulings. Nine of the disputes centered on food safety regulations, six cases involved technical regulations tied to customs requirements, and the remaining cases were in areas such as quotas, import bans, and disputes over environmental laws. It is important to note that although the majority of complaints brought to the WTO to date were from developed countries (20 cases), developing countries exercised their rights in seven instances. Developing countries also were respondents in five cases. Finally, disputes between developed members, such as the United States and the EU, bore obvious market-access implications for developing-country exporters. Two cases illustrate this situation and are next reviewed briefly.

The United States requested consultations with Japan on April 7, 1997, concerning the requirement imposed by Japan to test and confirm the efficacy of quarantine treatment for each variety of eight agricultural products: apples, cherries, peaches (including nectarines), walnuts, apricots, pears, plums and quince. Japanese regulation prohibits the importation of these products from the United States, on the grounds that they could potentially serve as hosts for the codling moth, a pest considered harmful to plant life. The import ban

TABLE 1.1. Provisions of the TBT and SPS Agreements Referenced in Consultations before the WTO Dispute Settlement Body (1995-2000)

Case	Measure	Complainant*	SPS	TBT
DS2	U.S.: Gasoline Standards	Venezuela		2
DS3	Korea: Testing and Inspection Requirements	United States	2,5	5, 6
DS4	U.S.: Gasoline Standards	Brazil		2
DS5	Korea: Shelf-Life Regulations - Frozen Meat	United States	2,5	2
DS12	EC: Trade Description Rules - Scallops	Peru		2,12
DS14	EC: Trade Description Rules - Scallops	Chile		2,12
DS18	Australia: Import Ban - Salmon	Canada	2,5	
DS20	Korea: Shelf-Life Regulations - Bottled Water	Canada	2,5	2
DS21	Australia: Import Ban - Salmon	United States	2,5	
DS26	EC: Import Ban - Hormone-Treated Beef	United States	2,3,5	2
DS41	Korea: Testing and Inspection Requirements	United States	2,5,8	2,5,6
DS48	EC: Import Ban - Hormone-Treated Beef	Canada	2,3,5	2
DS56	Argentina: Customs Duties - Textiles	United States		2
DS61	U.S. Ban on Shrimp Harvested by Nets	Philippines		2
DS72	EC: Exclusion From Tariff Schedule - Butter	New Zealand		2
DS76	Japan: Quarantine Regulations	United States	2,5,8	
DS85	U.S.: Changes to Rules of Origin Regarding Textiles	European Community		2
DS100	U.S.: USDA Decision on Poultry Product Safety	European Community	2-5,8, Ann. C	2, 5
DS133	Slovak Republic: Transportation Requirements	Switzerland	5	
DS134	EC: Import Duties - Rice	India	2	2
DS135	EC: Ban on Asbestos and Asbestos-Containing Products	Canada	2,3,5	2
DS137	EC: Import Restrictions- Wood of Conifers	Canada	2-6	2
DS144	U.S.: State Trucking Regulations Affecting Cattle and Swine	Canada	2-6,13, Ann. B,C	2,3,5,7
DS151	U.S.: Changes to Rules of Origin Regarding Textiles	European Community		2
DS203	Mexico: Anti-Dumping - Swine	United States	2,3,5,7,8	2,5
DS205	Egypt: Import Ban - Canned Tuna	Thailand	2,3,5, Ann. B	

*Does not include countries who subsequently requested to join consultations.
Source: Wilson, 1999.

could be lifted, however, if the exporting country proved that an alternative quarantine treatment achieved the required level of protection.

In 1987, Japan's Ministry of Agriculture, Forestry and Fisheries issued the varietal testing requirement at question in the dispute. The law prohibits the importation of products requiring quarantine treatment until testing has been performed on that particular variety. This is true even if the treatment has proven effective for other varieties of the same product. The United States maintained that these measures were inconsistent with Articles 2, 5, and 8 of the SPS Agreement. A WTO panel issued its report in October 1998, finding that Japan had acted inconsistently with Articles 2.2, 5.6, 7, and Annex C of the SPS Agreement. Japan appealed the panel's ruling.

In February 1999 the Appellate Body upheld the Panel's conclusion that Japan's varietal testing requirement, as it applied to apples, cherries, nectarines, and walnuts, was maintained without sufficient scientific evidence. The law was found inconsistent with Article 2 of the SPS Agreement. The Appellate report stated that there had to be a rational relationship between the SPS measure and the scientific evidence invoked to support the measure. It further noted that the existence of such a relationship had to be determined on a case-by-case basis, and would depend upon the particular circumstances of the case, including the characteristics of the measure and the quality and quantity of the evidence. It also concluded that the varietal testing requirement, as applied to apricots, pears, plums, and quince, was inconsistent with Article 5.1, noting that the risk assessment for codling moths conducted in 1996 did not refer to the requirement. In addition, it found that Japan had not reviewed any of its varietal testing requirements, as it was obligated to do.

In another case, Canada requested consultations with the European Community (EC) in May 1998 regarding a decree issued by the French government in 1996. The law banned imports of asbestos and of products containing asbestos. Canada contended that this import ban on asbestos violated articles 2, 3, and 5 of the SPS Agreement and Article 2 of the TBT Agreement. In addition, Canada claimed nullification and impairment of benefits accruing under these agreements. Resolution at the consultation stage proved unsuccessful and Canada requested that the Dispute Settlement Body establish a panel, which was formed in November 1998. The United States reserved its rights as a third party. The final panel report issued on July 25, 2000 upheld the ban by the EC rejecting the claim by Canada that the ban on asbestos constituted an unnecessary obstacle to trade.

The asbestos case is noteworthy for a number of reasons, including that the panel found the ban to be incompatible with the national treatment provisions of the General Agreement on Tariffs and Trade (GATT), but that the EC could apply the ban under GATT Article XX. This provision of the GATT states that nothing prevents members from imposing measures necessary to protect human, animal or plant life or health.

This was the first case brought to the WTO primarily on the basis of violations of the TBT Agreement and these findings raise important issues for the future debate on trade and product standards. Among these are whether intermediate inputs to final goods, such as asbestos used in construction materials, are covered under the TBT Agreement. The ruling that they are covered provides a wide scope of coverage for TBT disciplines in potential trade disputes on product standards. It also may establish a precedent for future environmental cases at the WTO, in which other exemptions for environmental standards that may restrict trade would be permitted under Article XX. In the case of asbestos, this material is increasingly used in developing countries for construction and intermediate inputs, such as brake pads for automobiles. If the EU ban were extended to global trade the implication for export markets could be grave.

It is noteworthy that most such disputes involve complaints by developed nations against practices of other developed nations. Thus, the industrialized countries have strong commercial, social, and environmental interests in managing the use of technical regulations affecting trade. The case involving the EU ban on imports of hormone-treated beef provides one clear example. The EU suffered an adverse WTO panel ruling in this case, based on the absence of scientific evidence to support its policy, but has yet to modify its import regime and continues to face significant retaliatory tariffs in the United States. On the horizon for global trade policy sits the issue of regulatory trade restraints against food products and industrial inputs that have been genetically modified. Given the substantial investments in such technologies already in place in the United States, Canada, China, and elsewhere, the potential stakes are significant.

In disputes over agricultural standards, developing countries have special difficulties and higher costs in demonstrating compliance with import regulations. As Finger and Schuler (1999) noted, the costs to developing nations of implementing requirements under the SPS Agreement in order to comply with obligations and exercise their rights are extremely high relative to feasible development budgets. Costs of implementing the TBT Agreement are likely equally high. Moreover, technology-intensive testing and certification systems are often needed to assure that products meet required standards, especially in areas of health and safety. Lacking this infrastructure poses clear problems for developing countries in meeting import requirements in conformity assessment and in defending their practices in WTO cases.

As noted above, the WTO experience indicates the need for empirical analysis to inform future deliberations. In areas where there are existing trade disciplines in the TBT and SPS Agreements, data on the quantitative trade effects of government regulations are lacking. A close review of cases brought to dispute settlement at the WTO, for example, reveals that little economic analysis was used to underpin panel decisions.

The WTO process would benefit from such economic analysis and empirical data. Basing trade-policy dialogue and decisions largely on administrative interpretations of trade law alone is increasingly problematic in areas of highly charged health, safety, and environmental laws. The WTO agreements require that regulations reference, where available, international norms and, in the case of the SPS Agreement, that regulations in areas of risk to health and safety be assessed on available scientific evidence. While this approach is valuable as a foundation for the rule of law in trade, as noted in Messerlin and Zarrouk (2000), reliance on scientific evidence alone to evaluate disputes may not provide adequate guidelines for their settlement.

This may be especially true in disputes over the use of environmental standards, eco-labeling requirements, and regulations for packaging and recycling. There is no international consensus on the net costs or benefits of these regulatory rules or their specific impact on exports of developing countries.[9] Some evidence suggests that labeling requirements pose significant obstacles for developing countries (Jha, Markandya, and Vossenaar 1999). Environmental standards and labeling are increasingly, however, a source of trade friction. For example, the EU has proposed mandatory labeling for packaging material to indicate whether the packaging is suitable for recycling. The regulation would differ from international standards and those in place in other markets. Informed analysis based on quantitative data would be beneficial for understanding the impact of such regulations and for building consensus and identifying specific implications of future trade-policy decisions in this area.

Disputes over policy and the lack of international consensus on how standards should be addressed in trade rules underpin the growing demand for deeper understanding of the relationship between standards and trade, both within the context of current "discriminatory" definitions in the WTO and "non-discriminatory" standards. Quantitative evidence on technical barriers would have particular relevance for developing countries. They must absorb information on international standards in order to meet their WTO obligations and face pressures to align national standards with international ones. Many developing countries, especially the least-developed nations, have limited resources available for participating in these processes. Moreover, there is no empirical evidence supporting the view that alignment or harmonization of regulatory standards always facilitates trade. Objective evidence would assist in understanding the precise nature of WTO obligations.

Developing-country members of the WTO consider standards and technical barriers a priority item in the post-Seattle trade agenda. A review of the formal and informal submissions by developing countries to the WTO in advance of the Seattle Ministerial in November 1999 reveals strong interest in addressing trade-related aspects of technical barriers and standards (Wilson 1999). This interest was expressed by large developing countries, such as India, high-income countries, such as Korea, and least- developed nations in Sub-Saharan Africa and Central America. These members raised two critical issues among many. First, they need technical assistance in implementing WTO obligations on standards and conformity assessment. Second, they are concerned about use of environmental regulations by developed countries to block imports. This interest was expressed by large developing countries, such as India, high-income countries, such as Korea, and least-developed nations in Sub-Saharan Africa and Central America. These members raised two critical issues among many. First, they need technical assistance in implementing WTO obligations on standards and conformity assessment. Second, they are concerned about use of environmental regulations by developed countries to block imports.

As China joins the WTO, it will be required to engineer deep reforms in domestic regulatory systems and import rules. This situation makes critical the development of new quantitative measures of the role of standards in trade if conflicts are to be avoided. A recent review of Ukraine's regulatory system commissioned by the World Bank, for example, reveals serious economic distortions in the way the government's standards, testing, and certification systems are designed (World Bank 2000a). In our judgment, reform of the standards system in Ukraine and other developing countries should be viewed as a development priority. Additional information on best-practice standards and new empirical data on the trade impact of standards would help serve both this objective and the need to meet WTO requirements.

Finally, it is important to note the work of the WTO in the Second Triennial Review of the TBT Agreement. These regular reviews of implementation and operation of WTO agreements provide a platform for advancing rationalization.[10] This is particularly true in regard to developing countries, as the discussions in 2000 included a priority focus on developing-member needs and challenges. However, the work in these regular reviews, which address nontariff, technical barriers to trade, has taken place largely without benefit of an analytical framework based on empirical data. A stronger base of economic analysis is needed as the WTO agenda extends deeper into areas of domestic regulation enforced through border measures.

Standards in Bilateral and Regional Trade Agreements: A Focus on Harmonization and Mutual Recognition

The number of regional and bilateral trade agreements has risen rapidly over the past decade.[11] All of these include some effort to address technical barriers to trade. Examples include the Common Market for Eastern And Southern Africa (COMESA) in 1993, Mercosur in 1991 and the North American Free Trade Agreement (NAFTA) in 1994. A significant focal point in these negotiations centered on trade facilitation, advancing regulatory best practices, and removal of technical barriers to imports.

In particular, the use of mutual recognition of testing and certification procedures and related efforts to harmonize standards are among the most important policy tools advanced in regional trade talks.[12] An important factor underpinning the legitimacy of MRAs is that the multilateral TBT Agreement encourages WTO members to enter into such arrangements. A total of 29 bilateral and regional MRAs and other memoranda of understanding on mutual recognition have been notified to the WTO. These types of arrangements have proliferated with little empirical analysis on the costs and benefits of alternative policy options. Discussions launched in 1994 within APEC, and forceful movement by the European Commission to conclude bilateral agreements with trade partners in the G-7, demonstrate anew the importance of standards and the need for empirical work to inform decision making.

The APEC Agenda

Among the most important regional trading arrangements involving both developed and developing countries, APEC represents the largest regional grouping formed to date. Members represent about 24 percent of world trade. The APEC goal to achieve open trade in the Asia Pacific region by 2010 includes an emphasis on trade liberalization and facilitation. This part of the APEC work program has assumed greater importance over time, in part because consensus agreement on lowered tariff barriers and investment liberalization has proven difficult to achieve. The APEC agenda on standards and technical barriers includes obligations under both Collective Action Plans (CAPs) and Individual Action Plans (IAPs). For example, APEC has launched programs to increase transparency in regulatory systems and exchange information on safety of toys. It has also funded projects such as training of customs officers in food safety inspection techniques.

APEC is driving broader policy discussions on standards and trade, in part through its relationships to sub-regional trade arrangements in the Asia-Pacific region. There have been 16 sub-regional arrangements in the APEC region since 1998, many of which have taken up the policy objectives APEC framed in standards in the early 1990s (Scollay and Gilbert 2000). An important objective in APEC is to move toward MRAs in regulated product sectors and related alignment (harmonization) of mandatory and voluntary standards with international ones.[13] This work is led by a Standards and Conformance Subcommittee in APEC. There is an MRA in place for exchange of information on toy safety, one on conformity assessment for food products, and an MRA under discussion covering electrical equipment.

Negotiating plurilateral MRAs such as these has consumed political energy in APEC and absorbed scarce government and private resources, with little evidence of concrete outcomes. Moreover, there are serious questions about the implementation of mutual recognition in APEC. The group acts on a consensus basis. Members can sign and implement MRAs among a subset of APEC nations and not extend privileges to others. Participation in these agreements by developing-country members is problematic, given their lack of technical parity in testing and certification infrastructure. Moreover, we know little about the net benefit of aligning national standards with international ones, especially for developing nations. Does harmonization benefit developing countries, even if they were not direct participants and lack the resources necessary to develop or absorb technology and information in standards? Significant pressure remains within APEC to harmonize standards in sectors such as electrical equipment, food labeling, and machinery.

The European Union: Exporting the MRA Model

The European Commission in the 1980s concluded several bilateral MRAs with other developed countries. The MRA model adopted is based largely on the framework developed in Europe to forge internal market harmonization. The EU has concluded MRAs with the United States, Canada, Australia, New Zealand, Switzerland, and Israel and is finalizing negotiations with Japan.[14] Product coverage varies with each agreement. The majority of these, however, cover regulations on products subject to electrical safety and electromagnetic compatibility requirements (radios, refrigerators, and the like), medical products, and telecommunications equipment, among others. The EU-Australia MRA includes motor vehicles and parts. The MRA with the United States is estimated by the European Commission to cover approximately $41 billion in bilateral trade.[15] Over $1.7 billion in trade in autos and parts is covered under the MRA with Australia.

Although the European Commission has moderated its position of full support for MRAs (Meyer 1999), it continues to negotiate them with countries seeking to join the EU and is considering extending them to other nations. The EU represents 22 percent of world trade (Bergsten 1999) and exercises significant influence in the developing world, especially in Africa, South

America, and countries in the Middle East and North Africa (MENA) region. The continued advocacy by the EU of MRAs as part of its trade agenda provides powerful incentives for countries to embrace this model. Is the MRA framework suitable for development objectives and trade liberalization? Based on existing anecdotal evidence and lack of empirical data, one would certainly question the utility of such a tool. The next section expands on this theme to illustrate directions for empirical research on standards and trade.

What Does the Trade Policy Context Suggest Are Priority Areas for Research?

We draw several conclusions from this review of standards in the trade-policy context for undertaking empirical research. As tariffs are eliminated, border requirements that might discriminate against foreign producers must be understood and addressed. Two issues are critical. First, information is needed in areas where government control is exercised through mandatory regulations on imports, thereby affecting the exporting firm's ability to produce to particular technical specifications or meet testing and certification requirements. Second, research can fruitfully focus on identifying the impact of barriers defined by WTO obligations as discriminatory. This approach would avoid the problems of attempting to assess the trade impact of all standards that might affect competitive positions in the marketplace. Priority work should also be placed on assessing the impact of domestic regulations on trade in those areas where debate continues about the boundaries of WTO disciplines in the TBT and SPS Agreements. Among the latter issues are environmental standards tied to eco-labeling requirements and the issue of what constitutes an "international" standard, as specified in the TBT Agreement. The developing countries have great interests in both areas, as well.

For the most part, the disputes taken to the WTO since 1995 under provisions of the TBT and SPS Agreements meet the test of clearly defined regulatory control exercised by government. None of the cases involved voluntary standards reached through market competition without regulatory intervention. Research and data-gathering efforts focused on determining the costs of divergent government health and safety regulations seem appropriate, emphasizing the meaning of discrimination within WTO rules. Determining whether regulatory harmonization and MRAs are viewed as important objectives by private firms could also shape the empirical research agenda. Examining the success or failure by developing-country firms in entering major export markets where MRAs are complete or under negotiation would provide valuable data to inform policy.

Finally, a focus on understanding the impact of regulation on trade in both goods and agricultural products is important. There is a modest analytical base in the literature on the trade effects of sanitary and phytosanitary standards to build upon, as outlined in Henson, Loader and Swinbank (1999) and Roberts (1999), among others. What is needed in this area is a richer base of empirical data on a global basis, especially to inform questions about the impact of standards in new regulatory areas, such as Genetically Modified Organisms. Firm-level data on costs of meeting differing sanitary requirements in the domestic market and those applied in major export markets are also needed, especially to inform debate over what constitutes "discrimination" in future trade talks in agriculture.

In regard to trade in goods, empirical research should help inform questions in which developing countries have a direct stake. As global trade has expanded, the composition of trade has been altered. Developing countries produce and market increasingly more manufactures as a share of total exports. Given the nature of global manufacturing, these countries are the primary location for production of components manufactured as inputs into final products. As regulatory requirements come into conflict and the dispersion of economic activity around the world continues, pressure on the WTO system to address technical barriers to trade will increase. Data on the precise impact of regulations on trade will help in a significant way toward a more informed public policy dialogue.

Analytical Overview of the Role of Standards in International Trade

A regulation is often defined as a mandatory requirement imposed by public authorities, whereas a standard is defined as a voluntary specification emanating from market forces (Sykes, 1995). Thus, competitors must comply with a regulation but may choose not to comply with a standard. While this distinction is useful for policy purposes, in this section we use the term standards to refer to both mandatory requirements and voluntary specifications.

Standards may be categorized first by function. Product standards refer to characteristics that goods must possess, such as minimum nutrition content, maximum toxicity or noxious emissions, performance requirements, or interoperability with components systems or networks. Production process or methods standards refer to conditions under which products are made. Such standards may be directly related to the product itself, such as a prescribed chemical formulation or rules of production for wines within a geographical area. They may be more broadly aimed at production conditions that are not directly related to the final good, such as working conditions. A further type of standard is a labeling requirement, which mandates provision of information to consumers about product characteristics or conditions of production.

The Need for Standards

Standards arise for numerous reasons. In principle, they are designed to facilitate production and exchange, reduce transaction costs, guarantee quality, and achieve the provision of public goods. They may also operate, by design or by circumstance, to restrain competition.

In the first instance, standards are demanded because they contribute to the provision of public goods. Emissions standards and fuel-economy requirements can contribute to cleaner air, though whether they are the most efficient instruments for this purpose is another question. Health and sanitation requirements can raise average health status in an economy, with spillover benefits into higher productivity. Interoperability standards buttress gains from information networks.

It is possible to think of standards as public goods themselves in that a standard must appeal to a group facing a similar problem, generating joint consumption benefits (Casella, 1996; Kindleberger, 1983). In that context, standards may be non-rival in consumption. However, they are not necessarily non-excludable, for market and technical mechanisms may suffice to provide a sufficient return to standards development. Thus, an important issue arises as to whether purely private provision of standards is sufficient and under what circumstances.

Because standards are tools for satisfying preferences, they do not remain constant over time, nor are they identical across countries. Demands for public goods vary with income levels, relative endowments of factors, information, technologies, and other variables. Accordingly, the derived demand for standards may be quite different in various countries.

Consider the potential market difficulties that standards might be used to overcome. First, standards can improve information flows between suppliers and consumers regarding the inherent characteristics and quality of products, thereby facilitating market transactions. For example, failing adequate information about products, consumers are vulnerable to the "lemons problem" (Akerlof, 1970). More generally, standardization may reduce the costs of uncertainty that consumers face in assessing product quality (Jones and Hudson, 1996). Costs include the time and effort consumers devote to search. Standards facilitate comparisons by consumers across products with common essential characteristics. Standards can increase demand for complementary goods. Users may mix and match components within a system, thereby permitting them to choose based on broader characteristics. In consequence, demand for both systems and components may rise.

Standards can also raise the elasticity of substitution in demand between versions of similar products (Harrison, et al, 1996). Because essential characteristics are standardized and quality and performance are guaranteed, products become closer substitutes. An important implication is that trade liberalization generates a more elastic increase in demand for imported goods under standardization than under non-standardization.[16]

By permitting producers to settle on a limited range of product characteristics or processes, standards and regulations can promote economies of scale (Stephenson, 1997). Sectors that had been segmented by variable standards can be rationalized by greater output scale, albeit at the potential cost of reduced product variety. Standards may provide focal points around which firms can organize their production processes. For example, the ability of enterprises to interchange inputs can reduce inventory costs and raise flexibility. Moreover, because intermediates may be subject to quality or performance standards, characteristics of final goods attain greater certainty.

Technical standards serve as benchmarks for technological capability and guarantees of compatibility with other components or with networks (OECD, 1999; David and Greenstein, 1990). Thus, countries may choose to keep their telecommunications or information networks open to entry of new devices and technologies, but would require such devices to operate effectively and safely within the network.

In sum, the essential point of standards is to support market development and facilitate transactions. They may also promote integration with global markets. The impacts may be both static and dynamic in nature. Adopting standards can improve resource allocation and help diffuse technical information embodied in products and processes. Indeed, standards themselves may embody considerable information about technical knowledge. Adherence to recognized standards provides incentives for firms to upgrade the quality and reliability of their products to required levels.

Perhaps the extreme form of this process arises in network industries, where standards are needed to promote the growth of attached users. An externality arises in that the value to any user of connection with the network depends positively on the number of other users, suggesting that networks may be under-provided in private markets. Technical standards for interoperability with the network can overcome this difficulty.

Standards and regulations are aimed also at overcoming market failures associated with public goods, such as environmental protection, sanitation, and plant, animal, and human health. It is unlikely that an individual firm would absorb the costs of investing in such standards. While the firm might try to signal its investment through some distinguishing marks or marketing programs, other firms could free ride on the development costs.

Thus, in a wide variety of circumstances it is likely that the social marginal values of standards exceed their private marginal values. Public intervention may be required to establish appropriate standards and requirements in cases where the objective is attainment of public goods.[17] In other cases, coordinated private solutions may be available through standards-setting bodies.

Standards and Trade Problems

To an important extent these observations about the benefits of standards apply across borders. Thus, a country's standards signal its product characteristics and regulatory preferences to foreign consumers and suppliers. Trade could be promoted as an extension of the market-building impacts of standards. By adhering to compatibility requirements, countries can improve their integration with global information and telecommunication networks.

However, standards and technical regulations also impose costs that could restrain competition. While this statement holds for both domestic and international markets, it is useful to illustrate these costs in the context of international trade.

The most straightforward problem is that costs of complying with standards may be higher for foreign firms than for domestic firms, implicitly erecting a trade barrier. Compliance involves one-time costs of product re-design and building an administrative system. It also involves recurrent costs of maintaining quality control and testing and certification. Moreover, there may be indirect costs, such as reformulating the ingredients of a food product because of a requirement to list its nutrition characteristics. Thus, a rich menu of cost-raising possibilities exists in which varying standards can raise entry barriers (higher up-front costs) or diminish the ability to compete (higher marginal costs). A variant of this problem is that firms must decide whether to establish a costly platform design that may be modified slightly to accommodate particular markets, or to design a product initially solely for the home market but with costly modifications required for export. The former strategy is more common among larger enterprises, while the latter characterizes smaller firms (OECD, 1999). Thus, compliance costs can provide an advantage to large firms in global competition.

Costs may be distinguished also between meeting the precise technical regulations and verifying that regulations are met. The latter task is called conformity assessment, and it presents the largest potential technical barrier to trade. Governments in importing countries may refuse to recognize tests performed by exporting firms or their public authorities and may not accept conformity declarations. They may insist on performing their own inspections of exporter premises and on inspecting imported shipments. Thus, conformity assessment is vulnerable to bureaucratic non-transparency and susceptible to capture by domestic firms seeking protection. Time delays are particularly problematic for products subject to short technical life cycles. Moreover, the costs of uncertainty in complying with such procedures can reduce the willingness to compete.

Developing countries lag behind developed countries in their capacities for effective certification and accreditation of testing facilities (Stephenson, 1997). This situation has three important implications. First, developing countries find it difficult to develop adequate standards and reach MRAs with other nations. Second, they also have not integrated themselves through accepting test results from abroad (Wilson, 1995). Third, because authorities in developed countries may not have much trust in inspection procedures in developing countries, the former group are liable to collaborate on standards and MRAs that exclude the latter group. The potential for trade and investment diversion in this scenario seems extensive.

Variable international standards can serve to segment markets, raising market power and erecting entry barriers. For example, a simple requirement that product packaging or instructions be stated only in the importer's language makes it impossible for arbitragers to ship goods to higher-priced markets with different languages.[18] Markets would be better integrated under a rule permitting firms to use multiple languages.

It is evident that governments and firms could establish strategic standards that aim to achieve market closure, alter the terms of competition in favor of domestic firms, or improve the terms of trade (Matutes and Regibeau, 1996; Fischer and Serra, 2001). In one example, if costs of converting to a new network are large, a subgroup of countries might find it advantageous to form an exclusionary standards union, to the detriment of excluded countries (Gandal and Shy, 1999). Standards not only could achieve static market exclusion but also could strengthen dynamic market power.

However, some principles may be used to consider the role of standards in restraining trade. First, if a standard or its enforcement is purely cost-raising (e.g., through delays in inspection or arbitrary fees) it is inefficient and should be removed. Second, if a standard is set at a level that is stronger than needed to achieve a particular policy goal, it may have protectionist intent by virtue of reducing foreign profits at the expense of domestic profits. One mechanism for defining such a protectionist standard is to ask whether it exceeds the regulation a government would choose if all producers were domestic (Fischer and Serra, 2001).

Third, if a standard (whether mandated or voluntary) is discriminatory in application or effect between domestic and imported firms, the margin of discrimination could be viewed as unnecessary protection and removed.

Fourth, we could question whether a standard is chosen that is least disruptive to trade among available policies. And, finally, a standard might be considered protectionist if it mandates excessive caution in relation to reliable scientific measures of risk. Note that the principles in this paragraph are incorporated into the TBT and SPS agreements.

Standards Coordination

Standards vary naturally across countries because of different levels of development, technological capabilities, endowments, and preferences. International harmonization of technical standards is virtually impossible. However, some forms of coordination could expand market access for developing countries. We discussed earlier some of the practical difficulties in achieving coordination in the WTO and among particular groups of countries. Here we make a few analytical points.

Developing countries may wish to achieve more effective representation at the deliberations of international standards-setting bodies. The standards advanced are typically voluntary but could have exclusionary effect when set by a small set of national or industry interests. Developing countries should also closely examine the adoption of technical standards developed in industrial countries. From an economic efficiency perspective, there is no reason why developing countries need to construct domestic capacities for standards development in all industrial sectors. By moving toward more harmonized regulatory approaches, developing countries would directly absorb the technological knowledge inherent in standards and also promote inward technology transfer. We know of no empirical study that examines this hypothesis.

Finally, exporters, in theory, may gain from mutual recognition agreements among themselves and with developed countries. However, the successful use of such MRAs is dependent on building effective and competent domestic standards and procedures for conformity assessment and inspection. In this context, technical assistance would be beneficial. As noted above, caution should be exercised in aggressively seeking MRAs. While there may be some gains to regulatory competition among MRAs (OECD, 1999), it is not evident that such arrangements necessarily avoid races to the bottom. Moreover, an extensive web of bilateral and plurilateral MRAs raises risks of extensive discrimination.

Organization of the Volume

With this background in mind, we have organized the volume as follows. In the next chapter, Keith Maskus, John Wilson and Tsunehiro Otsuki discuss the empirical context for analyzing technical barriers to trade. It analyzes the existing empirical literature, drawing some lessons for purposes of specification. It then considers the difficult problems that emerge in trying to locate reliable data and measure the impacts of technical regulations. In some ways these difficulties are more severe than the well-known ones arising from measuring trade barriers. The chapter concludes by outlining possible survey work to be done and by setting out an econometric framework that will be applied to the data it generates.

The remainder of the book proceeds from theory to empirical application. Chapter 3 is a revised version of a paper Richard Baldwin presented at the Brookings Institution annual Trade Policy Forum in April, 2000.[19] Baldwin analyzes numerous aspects of the problems developing countries see in trying to deal with the regulations their exports must meet. He voices the concern that the issue of differential trust between developed nations and developing nations could lead the former group to establish preferential standards unions or MRAs covering technical barriers to trade. The latter countries would then feel compelled to adopt the former group's regulations to avoid trade discrimination, despite the fact that these regulations may not be otherwise appropriate to their needs.

In Chapter 4 Mattias Ganslandt and James Markusen devise a series of analytical models that capture various aspects of standards, such as their ability to increase product substitution and to operate differently on fixed costs versus marginal costs. They implement these ideas in computable general equilibrium (CGE) models of small dimensions in order to illuminate the welfare and cost tradeoffs they produce. While these models are not calibrated to actual data and the method of capturing the operation and effects of standards is only suggestive, they provide interesting theoretical perspective on regulations, competition, and trade. Of particular interest is their treatment of regulations as market-segmentation devices. The idea that variable international standards could prevent price arbitrage and have powerful impacts on competition and welfare is confirmed in their simulation analysis.

Neil Gandal, in Chapter 5, provides a perspective from the industrial-organization literature about impacts of compatibility standards on international trade. After a survey of the literature on compatibility and standardization he discusses several industries that could provide natural experiments for measuring the effects of such regulations on international trade. Basic empirical evidence suggests that the adoption of particular compatibility rules can have significant effects on trade and the development of industry.

Chantal Nielsen and Kym Anderson, in Chapter 6, analyze problems emerging for the global trading system from concerns in wealthy countries, particularly those in the EU, about the health and environmental consequences of foods containing GMOs. The trade restrictions emanating from these concerns go right to the heart of the SPS agreement in the WTO and could spark considerable commercial conflict. The authors develop a CGE model of global agricultural trade and use it to consider the output, trade, and welfare impacts of potential EU import bans and preference revelation through labeling. They first simulate the effects of GMO-inspired increases in agricultural productivity in countries that adopt the new technologies in oilseeds and cereal grains. The substantial output gains from this technological change would place considerable pressure on farmers in the EU. This pressure could be blunted by an import ban, with negative implications for EU consumers and adopting regions. Nielsen and Anderson find that a labeling program, if it effectively represents preferences in the EU, would generate a far smaller trade impact than an import ban.

In Chapter 7, T. Ademola Oyejide, E. Olawale Ogunkola, and Abiodun S. Bankole provide perspective on the potential impacts of SPS standards on the trade prospects of Sub-Saharan Africa (SSA). They provide an overview of such regulations affecting imports in the United States, the EU, and Japan, noting the importance of these markets for SSA food exports. Based on existing work and applications of new data, they argue that SPS rules in major export markets may have significant restraining effects on exports. They also discuss problems and empirical methods that could be used to investigate these issues.

In the final chapter, Hank Hilton and Arik Levinson estimate the stringency of state-level environmental regulations and impacts of these costs on FDI in the United States, both in the aggregate manufacturing sector and for pollution-intensive and non-pollution intensive sectors. Their main purpose is to describe econometric difficulties that arise in trying to develop measures of the strength of environmental standards, pointing out that such standards are endogenous to underlying economic variables and that their effective costs would depend on a country's industrial composition. Because of such problems, they advocate the collection of survey data from designated firms over a period of years in order to see how changes in environmental standards and costs could affect FDI, output, or trade. Indeed, a considerable contribution of the chapter is to advance methods for dealing with the significant measurement and econometric problems that would face any attempt to relate standards to costs and trade, as discussed more fully in the next chapter.

Concluding Remarks

The application of product regulations and standards is becoming increasingly contentious as an implicit non-tariff barrier to trade. Many developing countries express rising frustration with regulations that vary across their export markets, require duplicative conformity procedures, and seem to change capriciously in order to exclude imports. Some developed countries are adopting MRAs that may lower costs for their trading firms but could result in greater discrimination against countries that do not belong to them. Substantial anecdotal evidence suggests that compliance costs with regulations can be high in relation to the value of products, which could deter entry into export markets altogether.

In this context, it is important to obtain as much information as possible about the quantitative implications of standards and regulations for costs and trade prospects of firms in developing nations. The World Bank is undertaking such work, which will survey enterprises in developing and transition economies.[20] Data will be gathered on firm structures, output and export patterns, and fixed and marginal costs in addition to costs of complying with home and foreign regulations. These data should form the basis for econometric work on costs and trade dynamics, the results of which should be informative for analysts and policymakers interested in the international impacts of standards.

This volume provides an analytical framework for researchers working in the complex area of regulations and trade. As the authors demonstrate, there are considerable difficulties in devising adequate measures and in interpreting results of econometric and simulation analysis. However, the issues are so important, and our stock of knowledge so limited, that there should be greater rewards from undertaking this comprehensive task in a serious and rigorous fashion.

NOTES

1. This paper focuses on product or process standards related to goods and agricultural commodities. It does not address standards as they may be applied in the service sector or labor standards.

2. A number of recent studies assess changes in barriers to trade in non-tariff areas, including those of particular importance to developing countries. See Hertel and Martin (1999) and Hertel, Hoekman, and Martin (2000).

3. For an overview of the status of implementation of Uruguay Round commitments of particular importance to developing countries see Finger and Schuler (1999). Wilson (2000) outlines specific implementation difficulties faced by developing countries in the Agreement on Technical Barriers to Trade. See also "Technical Barriers to Trade: Challenges and Opportunities for Developing Countries," Statement of the World Bank to the Technical Barriers to Trade Committee, World Trade Organization, Geneva, 24 February 2000.

4. For a comprehensive discussion of these procedures see National Research Council (1995).

5. For additional information on APEC see, www.apecsec.org and for TABD see www.tabd.org.
6. See Baldwin's chapter in this volume for further discussion.
7. An overview of all WTO dispute settlement cases may be accessed via the WTO website at www.wto.org/wto/dispute/bulletin.htm. and through the WTO Document Distribution Facility at www.wto.org/ddf. The U.S.-EU case on hormone treated beef is cataloged under WT/DS26 & WT/DS48.
8. See Wilson (1999) and www.wto.org.
9. For a discussion of the specific links between the TBT Agreement, eco-labeling programs, and trade see Chang, (1997).
10. Results of the first triennial review of the TBT Agreement are summarized at: http://www.wto.org/wto/goods/tbtindex.htm.
11. There is an extensive literature on the impact of regional and bilateral trade agreements on the world trading system. See for example; World Bank (2000b), Baldwin and Venables (1997), Frankel (1997), and Winters (1999).
12. The literature on MRAs from a trade policy perspective is limited. Among the most informative collection of papers on the subject can be found in OECD publications, including chapters by Roessler, Neven, and Nicoladis, among others in OECD (1996).
13. For an overview of the APEC agenda in standards see; www.apecsec.org. The "Report on Progress of Alignment with International Standards in APEC Priority Areas, APEC Subcommittee on Standards and Conformance, 1999" provides a status report on progress in alignment. Wilson (1995) includes a critique of the standards agenda, as based on member commitments in the Osaka Action agenda of 1995, which set the framework for current talks.
14. Taking a broad definition of MRAs to include binding cooperative agreements between government standards and regulatory agencies to share data, information, and other expertise, the EU has entered into 12 of the 29 total agreements notified to the WTO cited above.
15. The Transatlantic Business Dialogue (TABD) has provided a CEO-level platform providing analysis and advice on U.S.-EU trade policy since 1995. Detailed industry sector information on MRAs is found at www.tabd.org.
16. See Baldwin's chapter in this volume.
17. Standards may not be the most efficient mechanism available for this purpose. For example, effluent fees or emission permits may be expected to achieve a target reduction in emissions at lower cost than mandated technical specifications (Perroni and Wigle, 1999).
18. In their chapter, Ganslandt and Markusen simulate the costs of preventing arbitrage through the erection of standards.
19. The Brookings Institution has graciously agreed to permit us to include the paper in this volume.
20. More information is available. Please access the World Bank website at www.worldbank.org./research/trade.

REFERENCES

Akerlof, G. 1970. The Market for Lemons: Qualitative Uncertainty and the Market Mechanism. *Quarterly Journal of Economics.* 84:488-500.

Casella, A. 1996. Free Trade and Evolving Standards, in *Fair Trade and Harmonization: Prerequisites for Free Trade?,* ed. J. N. Bhagwati and R.E. Hudec, Cambridge, MA: MIT Press.

Chang, S.W. 1997. GATTing a Green Trade Barrier: Eco-Labelling and the WTO Agreement on Technical Barriers to Trade. *Journal of World Trade* 31: 137-159.

David, P.A. and S. Greenstein. 1990. The Economics of Compatibility Standards: An Introduction to Recent Research. *Economics of Innovation and New Technologies.* 1:3-41.

Finger, M., and P. Schuler. 1999. Implementation of Uruguay Round Commitments: The Development Challenge. Washington D.C.: World Bank.

Fischer, R. and P. Serra. 2000. Standards and Protection. *Journal of International Economics.* Amsterdam: Elsevier.

Gandal, N. and O. Shy. 1999. Standardization Policy and International Trade. Manuscript.

Henson, S., R. Loader, and A. Swinbank. 1999. The Impact of Sanitary and Phytosanitary Measures on Developing Country Exports of Agricultural and Food Products. Paper prepared for *Agriculture and the New Trade Agenda from a Development Perspective: Interests and Options in the WTO 2000 Negotiations.* Geneva: World Bank.

Hertel, T.W., K. Anderson, J. Francois, and W. Martin. 2000. Agriculture and Non-Agricultural Liberalization in the Millennium Round. *WTO 2000 Negotiations.* Geneva: World Trade Organization.

Hertel, T.W., B.M. Hoekman, and W. Martin. 2000. Developing Countries and a New Round of WTO Negotiations. Washington D.C.: World Bank.

Hertel, T.W., and W. Martin. 2000. Developing Country Interests in Liberalizing Manufactures Trade. *Background research for Seattle Ministerial Meeting.* Washington, D.C.: World Bank.

Jha, V., A. Markandya, and R. Vossenaar. 1999. *Reconciling Trade and the Environment. Lessons from Case Studies in Developing Countries.* London: Edward Elgar Publishing Co.

Jones, P. and J. Hudson. 1996. Standardization and the Costs of Assessing Quality. *European Journal of Political Economy* 12: 355-361.

Kindleberger, C. 1983. Standards as Public, Collective and Private Goods. *Kyklos* 36:377-396.

Matutes, C. and P. Regibeau. 1996. A Selective Review of the Economics of Standardization: Entry Deterrence, Technological Progress and International Competition. *European Journal of Political Economy* 12:183-209.

National Research Council. 1995. *Standards, Conformity Assessment, and Trade: Into the 21st Century.* Washington, D.C.: National Academy Press.

Organization for Economic Development and Cooperation. 1996. *Regionalism and Its Place in the Multilateral Trading System.* Paris: OECD.

Pardo Quintillan, S. 1999. Free Trade, Public Health Protection and Consumer Information in the European and WTO Context -- Hormone-Treated Beef and Genetically Modified Organisms. *Journal of World Trade* 33: 147-197.

Perroni, C. and R. Wigle. 1999. International Process Standards and North-South Trade. *Review of Development Economics* 3:11-26.

Roberts, Donna. 1999. Analyzing Technical Trade Barriers in Agricultural Markets: Challenges and Priorities. *Agribusiness* 15: 335-54.

Scollay, R. and J. Gilbert. 2000. *New Sub-Regional Trading Arrangements in the Asia Pacific.* Washington, D.C.: Institute for International Economics.

Stephenson, S. M. 1997. *Standards and Conformity Assessment as Nontariff Barriers to Trade.* Policy Research Working Paper no. 1826, Washington D.C.: World Bank.

Sykes, A.O. 1995. *Product Standards for Internationally Integrated Goods Markets.* Washington D.C.: Brookings Institution.

Wilson, J.S. 1995. *Standards and APEC: An Action Agenda.* Washington D.C.: Institute for International Economics.

———. 1999. "The Post-Seattle Agenda of the WTO in Standards and Technical Barriers to Trade: Issues for the Developing Countries," Washington D.C.: World Bank.

———. 2000. Technical Barriers to Trade and Standards, Challenges and Opportunities for Developing Countries Presented by the World Bank. Submitted to *the Technical Barriers to Trade Committee Meeting World Trade Organization.*

World Bank. 2000a. Draft Report on Standardization and Conformity Assessment in the Ukraine. Washington D.C.: World Bank.

———. 2000b. *Trade Blocs.* Washington D.C.: World Bank.

CHAPTER 2

An Empirical Framework for Analyzing Technical Regulations and Trade

Keith E. Maskus
Tsunehiro Otsuki
John S. Wilson

Introduction

The studies in this volume demonstrate that standards are complex in intention and operation. Economic theory is ambiguous in its predictions about how technical regulations affect trade. For example, propagating standards could assist a country's export performance by guaranteeing quality and performance characteristics. It could also raise import demand, further expanding trade. However, trade could be restricted if there is a cost-raising and discriminatory element to standards. Moreover, if standards expand scale but reduce variety the impact on trade would be unclear. Thus, the key relationships are ambiguous and require empirical study.

In this chapter we lay out the approaches that the research team at the World Bank will employ to study these relationships. The next section provides a review of the limited empirical literature available and concludes that considerable room remains for further analysis. Section 3 is devoted to problems of measurement that arise in a study of standards and trade. We discuss data collection in the Section 4. In the fifth section we set out a framework for econometric estimation of the trade, cost and productivity impacts of regulations, noting several problems associated with such estimation. We further discuss potential models for performing partial-equilibrium and general-equilibrium simulation analysis of standards and trade.

Review of the Empirical Literature

Four approaches exist in the literature for studying standards, regulations, and trade. These include surveys of firms' cost responses to regulations, macro-level econometric analysis of standards and trade, partial equilibrium models, and computable general equilibrium (CGE) models. Each approach provides useful insights and a basis for building further work.

Surveys

OECD (1999) conducted a survey of 55 firms in three sectors in the United States, Japan, the United Kingdom, and Germany. The sectors were terminal telecommunications equipment, dairy products, and automotive components. The purpose of the study was to examine the extent to which technical standards and conformity assessment procedures impede trade. Perhaps the most striking finding was that few firms considered standards to be of great concern. In telecommunications equipment firms reported wide differences in standards but most thought these differences amounted only to a minor problem. In dairy products there were problems in certification and approval delays for exporters of specialty products, but dealers in bulk dairy goods reported few difficulties. In auto parts there were significant costs imposed on a small range of products.

Among those firms that estimated their additional costs of complying with foreign standards, the estimates ranged from no cost increase to ten percent, with most coming near the bottom of this range. A few firms reported that they chose not to enter a particular market because of expected regulatory costs. These firms tended to be the smaller ones in the sample, suggesting that standards do serve as relatively greater entry deterrents for small enterprises.

The interviews conducted by OECD supported a number of conclusions. First, harmonization can be helpful in lowering costs of product redesign and testing. Second, mutual recognition agreements (MRAs) on conformity assessment procedures reduce trade costs. Some respondents claimed that competition among approval agencies has lowered costs of application and registration. Third, meeting the terms of voluntary standards was a larger issue than meeting mandatory ones. In effect, mandatory standards certify that a product may be imported but do not provide entry into local voluntary codes and requirements. Fourth, small companies relied heavily on external sources of information and could not spread these costs among much output volume. They were, therefore, considerably less likely to export than were large firms.

The United States International Trade Commission (1998) performed informal interviews of corporate executives, officers of trade associations and government officials regarding the importance of standards as trade impediments in the information technology (IT) industry. Surveys were conducted in the United States, the European Union, and various countries in Asia and Latin America. The authors could provide only testimonial statements and did not attempt to categorize their results statistically.

Their principal finding was that many IT firms considered duplicative and discriminatory testing and certification requirements to be a substantial barrier to trade. Indeed, some representatives of the U.S. industry claimed that standards-related costs mounted the most significant trade restriction in the sector. One reason is that satisfying multiple conformity assessment procedures and varying labeling requirements in different countries is costly because inputs and products are sourced from numerous locations. This makes it difficult to apply varying inspection rules and labels to products, the ultimate destination of which is unknown at time of production.

A further cost is that approval and testing procedures in importing nations are often lengthy in duration, a substantial problem in a sector with short average product life cycles. For example, meeting the EU's tests for telecommunications equipment was estimated to take six to eight weeks, reducing product value by five to ten percent (USITC, 1998, 4-2). Similarly, costs of registering products to recognized standards, such as ISO 9000, exceeded $245,000 per American telecommunications firm. These firms thought the registration was unnecessary and redundant in view of their own quality assurance systems. An estimate by an industry association claimed that duplication of mandatory U.S. and EU testing for computers, telecommunications equipment, and similar IT products raised costs for U.S. companies by more than $1.3 billion annually (USITC, 1998, 4-8). Finally, some importing nations require compliance with specific norms that duplicate international standards, raising an additional and unnecessary testing cost.

The study noted that some industry analysts thought the 1997 U.S.-EU MRA in IT products (and other goods) would reduce direct costs to IT firms by $1.3 billion and would achieve further gains through shrinking approval periods and raising competition among testing laboratories (USITC, 1998, 5-3). American trade officials thought that the 1998 APEC MRA on telecommunications equipment would increase intra-APEC trade in IT goods, which amounted to $45 billion that year. Finally, some analysts noted that potential efficiency gains could be realized if governments shifted toward unilateral recognition of foreign conformity assessment and supplier's declarations of conformity. Henson, et al (2000) studied problems that developing countries have in meeting the SPS requirements of the developed countries and in adhering to the provisions of the SPS Agreement. They interviewed officials of multilateral organizations and performed a survey of developing-country members of the WTO or Codex Alimentarius. They claimed that developing countries are strongly constrained in their ability to export food products by SPS mandates in developed nations. Such requirements were ranked as the most significant constraint on exporting agricultural and food products to the EU, ranking ahead of transport costs, tariffs, and quotas. Among destination markets, the European Union afforded the most problems, followed by Australia, United States, Japan, and Canada.

Unfortunately, survey questions were qualitative in nature and could support no statistical analysis. The authors illustrated potentially costly impacts on developing-country exporters through a series of examples of SPS measures. These case studies provide an interesting compendium of circumstances under which exporters might find foreign market access restricted by a failure to maintain sanitary conditions in production or through hasty decisions by importing nations. The report also detailed difficulties that developing countries face in effectively participating in the SPS Agreement.

Econometric Studies

Two recent studies have related trade flows to measures of a country's stock of standards. Both Swann, et al (1996) and Moenius (1999) discussed the multiplicity of economic hypotheses about trade and standards, noting that virtually any prediction could be supported. Swann, et al (1996) regressed British net exports, exports, and imports over the period 1985-1991 on counts of voluntary national ("idiosyncratic") and international standards recognized by the United Kingdom and Germany. Standards counts were taken from the PERINORM database and concorded to the 3-digit Standard Industrial Classification (SIC). There are numerous problems with using counts of standards to measure the effective stock of technical specifications that could affect trade.[1] Standards vary in importance across sectors and products and different standards would not be expected to have similar effects.

The authors discovered that British national standards tended to raise both imports and exports. The coefficient in the imports equation was 0.34, while that in the exports equation was 0.48. This suggested that a 100-unit increase in British national standards would raise manufacturing imports by 34% and manufacturing exports by 48%, which are highly elastic responses.[2] International standards to which Britain was a party had little impact on imports but a positively significant effect on exports, though at 24% this effect was smaller than that of the national standards. German standards significantly raised British imports and showed negative effects on British exports.

The authors concluded that the imposition of British standards had a positive effect on both UK exports and imports. Thus, they served both to signal quality abroad and to raise import demand in the UK. They interpret the positive effect in imports to imply that idiosyncratic British standards raise costs disproportionately for domestic firms, bringing in lower-cost imports. It is impossible to make such a fine inference from their work, however, for the theory and regressions do not support such distinctions. They further concluded that German standards reduce British exports, perhaps suggesting a protective impact. Note that the two findings are inconsistent and the authors offered no explanation. Finally, they found that idiosyncratic standards raised trade more than international ones. They posited that the smaller effect of international standards reflected a tradeoff between more trade based on higher economies of scale and less trade associated with reduced product variety. This hypothesis went untested, however.

Moenius (1999) advanced this approach by incorporating standards counts data from PERINORM for 12 countries and 471 4-digit industries from the Standard International Trade Classification (SITC) into a gravity-based analysis of bilateral trade volumes over the period 1980-1995. He distinguished between country-specific standards, measured by the number of documents specifying a technical requirement solely within a country, and bilaterally shared standards, measured by the number of documents linked between two countries covering the same code. Objections raised about the information content of count data are particularly problematic with these linked standards. The sample used in the regression analysis included eight countries currently in the EU plus Switzerland, Australia, Poland, and Turkey, with U.S. and Japanese counts also made for comparison purposes. Note that this is the only study to incorporate data from a transition economy and a developing economy.

At the simplest level, it could be hypothesized that country-specific standards are implicit non-tariff barriers and restrict trade flows, while shared standards remove the underlying cost differences and raise trade. However, there are many channels through which both standards types could affect international trade. For discussion purposes it is useful to reproduce his summary table here as Table 2.1. For any geographical designation of standards, hypotheses exist that trade could either rise or fall. However, there seems a presumption that shared standards should increase both exports and imports on balance, importer-specific standards should reduce imports, and exporter-specific standards should have indeterminate effects on exports. Note that the sign pattern in each row is distinctive, so that econometric analysis bears some promise of distinguishing among these theories.

Moenius (1999) regressed bilateral trade volumes in 4-digit SITC sectors on counts of shared standards and an array of dummy variables to account for other trade determinants in country pairs and years. He found that shared standards had a positive and highly significant effect on trade volumes, with a one-percent rise in shared standards associated with perhaps an 0.32 percent increase in trade. This result held up when he added a lagged dependent variable to control for first-order serial correlation. By his calculations, taking an average elasticity across his specifications of 0.35, a one-percent rise in shared standards between the United States and its trading partners would raise U.S. trade volumes by $6 billion, which seems economically significant. Finally, causality tests could not reject the hypothesis that increases in standards raise trade volumes and increases in trade generate higher numbers of standards in a feedback effect.

He also regressed bilateral import volumes on the same variables but added importer-specific and exporter-specific standards, along with dummy variables selecting 2-digit aggregate industry categories in an attempt to control for relative price variations across sectors. He found that shared standards raise imports significantly, with an elasticity of 0.16. Interestingly, the number of importer-specific standards also tended to increase imports slightly, rather than to reduce them as expected if standards are NTBs. Exporter-specific standards had a strongly positive impact on import volumes, with an elasticity of perhaps 0.27, supporting the idea that such standards provide a signal of quality and reliability to importers.

Moenius refined this result by repeating the analysis for industries grouped at the one-digit SITC level. He found that importer-specific standards significantly reduced imports for the non-manufacturing sectors but significantly raised them for manufacturing sectors. In contrast, exporter-specific standards were positively associated with most grouped trade flows. This is his key finding and the author advanced an intriguing explanation for it. In particular, standards and codes have offsetting impacts on costs. By forcing adaptation, testing, and certification to meet needs in particular markets, importer-specific standards raise compliance costs and should reduce trade. However, both shared standards and country-specific standards should reduce the costs of acquiring information about market preferences and product

TABLE 2.1. Predictions of Theoretical Literature about Standards and Trade

Theory	Shared Standards	Country-specific Standards Importer	Country-specific Standards Exporter
Non-Tariff barriers:			
Mainstream/strategic alliances	+	-	
Competitive disadvantage		+	-
Standardization trap	+	-	-
Competitive Advantage		-	+
Loss of Variety	-	+	+

Source: Moenius (1999).

quality, serving an important signaling function and raising trade.[3] The reduction in information costs should be particularly valuable for manufacturing products, which are subject to more variety in characteristics and technological specifications. The cost-raising aspects of standards may be more prevalent for more homogeneous non-manufactures.

Otsuki, et al (2000) estimated a gravity model of trade in agricultural products between 15 European countries and nine African countries between 1989 and 1998. Their results suggested that the implementation of a new aflotoxin standard in the EU would reduce African exports of cereal, dried fruits, and nuts by 64 percent, or US $ 670 million, in relation to what trade would be in the presence of an international standard. This finding suggests that the impact of technical regulations could be particularly severe on agricultural exports from poor nations.

Partial Equilibrium Studies

Partial equilibrium models provide a framework for analyzing tariff-rate equivalents of standards and technical regulations and associated welfare changes. Tariff-rate equivalents are estimated as a price premium that exporters must overcome in addition to the world price plus tariff. Welfare change is estimated by investigating impacts on domestic consumer and producer surplus caused by an increase in costs to comply with standards and possibly also the change in domestic utility reflecting improved product quality. Demand and supply elasticities are often calibrated from existing studies. At the sacrifice of generality, the partial equilibrium approach has the advantages of transparency and comprehensiveness in analyzing changes in various welfare components and in incorporating specific details of standards and regulations.

Specification of the manner in which standards affect markets varies across studies, depending on whether regulations are thought to benefit consumers, producers, or both, and on whether exporters are obliged to face a higher implicit tariff rate or higher compliance costs. Thilmany and Barret (1997) studied the implications of technical regulations for dairy exports from the United States to the other NAFTA member countries. In their approach, both demand and supply curves were assumed to shift up. The shift in the demand curve reflects the effect of standards to alleviate consumer uncertainty about product quality, and the shift in the supply curve is due to increased transaction costs of export including compliance costs. While they were unable to separate non-tariff and tariff effects in their empirical analysis, they compared domestic and international prices to estimate the producer-subsidy equivalent and import tariff-rate equivalent of these trade barriers. They indicated that domestic producers are not necessarily subject to the standards, and that standards are often used to protect domestic producers as well as tariffs, thereby hurting domestic consumer welfare.

Paarlberg and Lee (1998) studied the case of U.S. tariff protection against beef imports from countries that may transmit Foot-to-Mouth Disease (FMD). In their approach, the domestic government is assumed to maximize the country's welfare by setting the optimal tariff rate, where expected loss of domestic beef production due to the FMD infection of U.S. livestock has been incorporated ex-ante into the tariff rate. They found that the magnitude of the optimal tariff, in the presence of an FMD risk, is very sensitive to the specification of the risk of importing FMD and to the size of output losses expected from an FMD outbreak. For example, holding expected output loss constant at 15.5%, the optimal tariff would rise from 8.1% to 929.4% when the import volume that is expected to cause one outbreak falls from 215,000 tons (low risk) to 24,000 tons (high risk). Put another way, an increase in the anticipated U.S. beef output loss from 1.4% to 15.5% would cause the tariff to rise by 86.8% and 929.4%, respectively, when one outbreak is expected for every 215 thousand tons of imports.

Calvin and Krissoff (1998) studied Japanese imports of U.S. Red and Golden Delicious apples by calculating price and welfare effects of sanitary and phytosanitary standards. They compared Japan's welfare under two scenarios, with and without a loss of domestic production due to transmission of fire blight disease through importation. The tariff-rate equivalent of the standards was estimated to be 27.2% in 1994-1997, which is high compared with actual tariff rate of 19.3%. When the production loss due to the disease outbreak was not considered, producer welfare loss directly related to trade from elimination of the standards was estimated to be $210 million per year, or 30% of original producer welfare. On net the welfare change in Japan was still positive, at $ 70.9 million, because there were larger gains in consumer welfare and tariff revenue. In the presence of production loss, the domestic supply curve was shifted up to reflect additional production costs such as pruning. It was found that a 26% domestic output loss would be required to offset the net trade gain from eliminating the standards. The statistical probability of such a loss would be negligible. One drawback of the analysis is their rather strong assumption that the difference between the Japan's domestic and world prices plus tariff rate are solely ascribed to the price premium for the standards.

Computable General Equilibrium Studies

Computable general equilibrium models provide considerable scope for understanding how changes in standards and technical regulations might affect trade and investment in various market settings. However, their measures of standards are necessarily heavily aggregated and cannot capture the complexities of codes as they exist at the detailed sectoral level. In that sense CGE studies incorporate crude specifications of standards into complex theoretical specifications, generating interesting predictions about how liberalization of technical barriers to trade could alter competitive prospects and rationalize industry. As policy guides, they are only suggestive.

Two studies are noteworthy in this context. Gasiorek, et al (1992) set out a CGE model of the EU economy with increasing returns to scale (IRTS) in several sectors. They modeled two scenarios, both of which assume that harmonization of standards in the EU would reduce trade costs by 2.5 percent. That is, technical regulations were assumed to be "sand in the wheels" and their harmonization would remove the associated inefficiencies. In the first case markets in various EU countries would remain segmented by other factors, permitting firms to mark up prices on the basis of local demand. In the second case markets would be fully integrated and firms would set uniform prices throughout the region based on EU-wide demand. They also distinguished between short-run effects, with fixed number of firms, and long-run effects, with endogenous firm exit.

This model generated large impacts on production and trade. In the IRTS sectors, EU production and exports would rise considerably due to the smaller costs. There would also be substantial trade diversion as imports from outside the EU were displaced by intra-regional production. Such impacts were even larger in the integrated-markets scenario. Welfare gains for the EU could be sizeable, at perhaps one percent of GDP in the short run and higher in the long run as inefficient firms leave production.

Harrison, et al (1996) extended this work in a number of ways. They questioned whether even full harmonization of standards would generate complete price integration in light of American evidence suggesting otherwise.[4] They advanced a CGE model with monopolistic competition in which the EU would gain from additional competition and rationalization but firms would retain a reduced ability to price discriminate. They adopted the Gasiorek, et al convention of a 2.5 percent reduction in standards costs. However, they added a demand-side channel for gains from harmonization. In particular, harmonized standards should raise information about foreign products and raise consumer confidence in their characteristics, raising the elasticity of substitution between home and other EU products. They implemented this notion by permitting the intra-EU substitution elasticity to rise from a level where other EU products compete with products from outside the EU to a level where all EU products are considered home goods.

They found that the 2.5 percent fall in standardization costs would generate a welfare gain of up to 0.5 percent of GDP. Adding the possibility of higher substitution would more than double this gain, albeit at the cost of shifting demand away from extra-EU imports. Finally, they calculated that in the long run investment would increase due to a rise in the real return to capital. In the new steady state equilibrium, welfare benefits could reach 2.4 percent of GDP per year.

Problems of Measurement

The prior discussion suggests that it is difficult to develop adequate measures of the stringency and effects of TBTs. Indeed, they have been characterized as "...one of the most difficult nontariff barriers imaginable to quantify" (Deardorff and Stern, 1998, p. 119). There are several conceptual and practical reasons for this difficulty, which we now review, with the aim of informing future analysis.

Basic Notions

Consider the simplest case of measuring the effects of a pure cost-increasing industrial standard that is imposed on imports coming into a market in Figure 2.1. In the diagram, ED represents the domestic country's excess-demand (import) curve, while ES depicts the foreign country's excess-supply (export) curve. Suppose the standard erects an additional cost of c dollars per unit imported, due to conformity assessment or inspection requirements, shifting ES up to ES + c. The following questions could be posed about the effects of this standard.[5]

Is the standard discriminatory? In the diagram, if ED did not incorporate an identical cost standard, the rule would discriminate against imports, causing them to fall as depicted and forcing exporters to absorb a lower net price, p_1^f as domestic price would rise to p_1^d. However, if the same standard were imposed on domestic production (here assumed to be perfectly substitutable for imports), the ED curve would rise by the same amount, with the effect being no change in quantity imported or exporter price but a domestic price that would be higher by c. In this context, a higher domestic price for the imported good is not necessarily indicative of a discriminatory standard and the usual tariff equivalent price measure of $(p^d-p^f)/p^f$ may be misleading unless compared to the cost increase imposed on domestic firms or to import quantity.

What are the appropriate measures of price and quantity change? In the simple case presented here, the compliance cost would generate an additive wedge between the foreign and domestic prices at any level of imports. However, the higher is the foreign price, the lower is this percentage tariff-equivalent, so that the standard price measure would depend on the final equilibrium point. This point is important when we recognize that the same absolute cost standard, if imposed by different importers with varying elasticities of excess demand, would generate different price measures of its restrictiveness. Put another way, quantity effects of the standard depend on market circumstances so it becomes difficult to assess them across countries without more information. A purer measure, independent of the demand curve, would be to compare prices at the unchanged quantity level, by defining the tariff equivalent as $(p_0^d - p_0)/p_0$. However, it is unlikely that these prices are observable unless one has time-series information on prices and imports just before and after a standard is imposed.

There are many practical reasons why such price-based measures could be flawed. It is not possible to use a measure of c to calculate the implied do-

FIGURE 2.1. Price and quantity effects of a technical standard

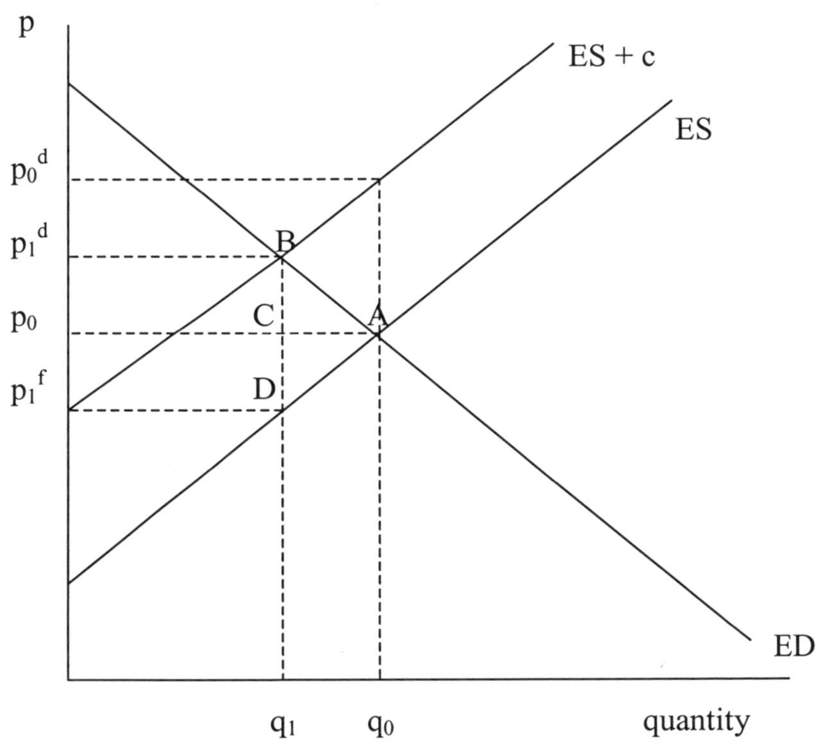

mestic price in the absence of such cost data. However, technical regulations are not specified as some absolute monetary cost per unit imported. Rather, they are rules under which inspection or conformity assessment procedures will be adequate for market entry. Thus, the per-unit cost would vary by exporting partner, depending on underlying costs, the efficiency of assessment mechanisms, and the initial volume of imports if there are economies of scale in testing. These issues are taken up further below.

What is the impact on the shape of the export supply and import demand curves? Again, regulations are unlikely to entail this simple additive cost. If they raised costs as a constant proportion of price or import value, the new ES curve would pivot upward from the intercept and be steeper than the original one. While the tariff-equivalent measure of the cost would be independent of quantity, the impact on imports still would depend on demand elasticity.

If a regulation raised fixed costs of exporting but had no impact on marginal costs, the proportionate cut in imports from small-volume exporters would be larger than that for large-volume exporters. It is possible that such a change could eliminate imports from smaller trading partners in the long run. Thus, if p_1^d were the domestic price of the most efficient exporter with the regulation in place, other exporters would need to meet that price, which could be impossible on a small volume with increasing returns to testing. It is evident that regulations can be discriminatory in effect if not in intention. In this regard, it is important for researchers to attempt to distinguish between impacts of technical regulations on fixed costs and variable costs.

The regulation could also affect the ED curve. If it were a requirement for customs inspection, for example, costs and delays could make it operate in a similar fashion as an import quota, with the demand curve becoming highly inelastic after some manageable level of imports. In this case, the tariff-equivalent price measure would depend on the proximity of import volumes to the inelastic range.

How does the regulation affect preferences? The position of the ED curve reflects a degree of consumer confidence in the imported good in the absence of a technical standard. The standard could signal consumers in the importing market that a product is safe, thereby reducing consumer uncertainty associated with inadequate information. The ED curve would shift to the right in Figure 2.1, reversing the decline in imports associated with the cost impact. It is possible over time to observe a higher domestic and foreign price and also a higher import volume. In this context, the original price wedge would become a misleading indicator of the impacts of the standard and it would be important to track market responses over time.

A further mechanism by which demand would be affected is that the regulation could increase the elasticity of substitution between the imported good and domestic products and products imported from other nations. How this would affect the ED curve would depend on the nature of substitution, as demonstrated in the Ganslandt-Markusen paper in this volume. Ordinarily, however, it would be expected to increase the price elasticity of import demand, rather than reduce it as is often be the case with more conventional NTBs.

Complications

These basic observations point to the inherent difficulty in sorting out the various impacts of a technical regulation on trade, prices, and ultimately economic welfare. However, additional complications arise that should be considered in an accounting of economic impacts. Many of these are discussed at length in other papers in the volume and we only highlight them here.

An obvious difficulty is that the impacts of regulations could vary considerably over time, depending on shifts in market demand and supply. Thus, for example, the tariff-equivalent price impact would depend on exporter price, itself a function of such variables as changes in technology, weather, and trade policies in third markets. Such variability can impose an additional cost on society.

One reason for regulating imports through sanitary and technical standards is to manage associated risks, such as an outbreak of a plant or animal disease. However, a regulation may not fully eliminate risk from imports; indeed, we would not anticipate an optimal regulation (one that balances costs and benefits at the margin) to do so. Rather, the regulation may reduce the probability of an outbreak and alter the variance of its distribution. Therefore, the import demand curve could shift in complex ways after the introduction of a standard. Imagine, for example, that in the absence of a regulation the probability of disease rises with the quantity of imports and is highly uncertain.[6] Further, suppose that the increase in marginal costs imposed on domestic producers in the presence of an outbreak depends on the quantity of imports but is uncertain. If consumers and producers were both risk-averse, a mean-reducing risk-management regulation would expand both consumption demand and domestic supply, with ambiguous effects on the ED curve in the neighborhood of free trade. However, if the regulation did not reduce the probability of an outbreak but tightened its variance, we would observe a rise in excess demand. Thus, the nature of the underlying risks and how they are affected by policy is of interest. It is evident that this factor would further complicate welfare calculations.

While technical barriers can manage risks, they may be subject to considerable uncertainty themselves. To the extent that they are applied in an arbitrary or variable manner, they would impose uncertainty costs on exporters. These costs could have potentially significant negative impacts on exports in the presence of fixed costs or an inability to divert products among destinations, and could limit investment in export facilities. This possibility seems particularly relevant in the area of technical regulations and standards, which may not be transparent due to their complexity.

In another vein, products are likely differentiated rather than perfect substitutes. In a differentiated-products model with varying marginal costs among suppliers, a common standard can have distinctive impacts across competitors. Thus, for example, Egyptian and Moroccan exporters might react differently to a regulation imposed in the United Kingdom. Alternatively, the reaction of an Egyptian exporter to a regulation in the United Kingdom might differ from its reaction to a similar regulation in Germany. Moreover, when products are differentiated a country could be both an importer and exporter of products affected by the same standard. In this case, a regulation that reduced the risk of importing a disease could promote exports by raising consumer confidence abroad. Tracing through such complicated effects would be a difficult empirical task.

Three other issues emerge in this area that could be important. First, markets may be imperfectly competitive, unlike that in Figure 2.1. It is evident that firm reactions to technical regulations could be markedly different in the case of oligopoly under varying degrees of increasing returns and entry conditions. Second, the effects of standards could depend on the existence of other standards, taxes, and market distortions. There may be interaction effects across sectors through inter-industry linkages or allocative decisions arising from standards, the extent of which would depend on alternative market barriers. These types of questions could be analyzed most readily in a CGE framework.

Finally, as some of the papers in this volume attest, standards could be endogenously set in some political economy process. Presumably such an outcome would be reflected in standards or procedures that discriminate against foreign suppliers, suggesting that a finding of extensive discrimination in could be evidence of endogeneity. It is possible that the trade-restrictive impacts of regulations could be underestimated if this causation is not accounted for. However, such an accounting lies beyond the scope of the current project.

Practical Difficulties

These conceptual problems are compounded by certain practical complications that make direct measurement of technical regulations especially difficult. First, they are complex in design and operation. Standards are not simple taxes, nor are they quantitative limits on trade or percentage procurement preferences. Rather, they are complicated specifications of such characteristics as minimum quality, maximum toxicity, ambient characteristics in the production environment, and technical compatibility, along with rules for demonstrating conformity. If products meet such rules they are permitted access without further limitation, at least in principle. Direct quantification of such rules typically is infeasible, except in cases where they are written as numerical limits and are comparable across countries or products. Rather, they need detailed description and micro-level analysis in which responses of individual firms are investigated.

A related issue is that the effects of standards may depend on their method of administration. It is evident that unreasonable delays in inspection and certification can erect a barrier to business certainty and trade. Sampling techniques and requirements could vary by trading partner, affecting export decisions. If administrative procedures are sufficiently unclear and variable, they could have a significant deterrent effect on investments in export facilities that would restrain trade more than the regulations themselves might suggest.[7]

A final practical difficulty relates to aggregation. Researchers may wish to characterize the regulations of particular countries, or facing particular products, in an index that captures some aspect of their operation, such as severity of trade impacts or degree of information revelation. In doing so, analysts would face the usual problems of deciding appropriate weights to choose in the aggregation. Import or export weights can be misleading if regulations limit or expand trade considerably. Production weights by importer or exporter would presumably, but not necessarily, be less biased. The best choice would be world production weights but such data may be difficult to find for detailed product categories.

The more difficult problem is that the complex nature of regulations in itself may make aggregation questionable. As noted below, simple counts of the number of regulations mixes those with significant effect and those with little impact, biasing the overall index. They also bunch together regulations of different types (quantity-based, quality-based, process requirements, inspection and certification rules), which may not be comparable. To make them comparable, analysts need to normalize them in terms of some economic impact, such as a relative cost or price effect. These economic impacts could then be weighted to establish the aggregated index. Even then analysts should state clearly what the information content of the index really is.

The list of conceptual and practical problems we have discussed pose daunting problems for empirical analysis. However, we contend that many of these issues may be addressed satisfactorily through firm-level survey analysis that permits careful and detailed investigation of how enterprises react to technical regulations. For example, the problems with the price-based measures noted above stem from the desire indirectly to infer the effects of regulations through associated price data. Such data may not be capable of identifying the supply-side effects of interest to exporters. Thus, surveys focused on cost impacts and actual trade decisions should be informative. In the following sections we present central aspects of the proposed research program.

Data Collection

Our ultimate objective is to develop useful evidence on the scope and consequences of TBTs for firms in developing countries. This requires specification of an empirical framework for the analysis. The initial task is to gather data that would improve our knowledge base. Among the many issues that could be investigated, the primary ones are as follows. First, within selected sectors and countries, how pervasive are regulations that could affect trade? Do exporting firms face more (or more costly) regulations imposed by domestic governments or by governments in importing countries? Does this mix of regulations vary by level of economic development? Second, what are the costs of complying with such regulations? These costs come in three categories: production or design costs, conformity assessment costs, and costs of meeting variable standards in different markets. Third, do mutual recognition agreements bear potential for reducing such costs?

These complex questions cannot be answered with currently available information and may defy precise quantification. Nonetheless, a comprehensive effort to gather and analyze data would create an important information base and improve our understanding of the relevant processes.

Policy and Industrial Surveys in Developing Countries

The World Bank research team will design and administer surveys of manufacturing firms in key developing nations. These instruments will elicit information on cost structures, production and exports, impediments to domestic sales and exports, and operations to comply with regulations. They should be administered to as many firms as possible, depending on budgetary costs. Public agencies and standards-setting bodies would also be surveyed to elicit information on the important standards and TBTs in each country and how these are established.

Country Selection. The countries included should cover a range of economic development and export experience yet have sufficiently deep agricultural and industrial structures to permit sectoral comparisons. The research team has chosen the following list of nations for study:

Sub-Saharan Africa: South Africa, Senegal, Mauritius.

Latin America and Caribbean: Argentina, Chile, Panama, Honduras.

South Asia: India.

East Asia: Thailand, Malaysia, Vietnam.

MENA: Egypt, Morocco.

Europe: Poland, Czech Republic, Bulgaria.

Sector Selection. The sectors chosen should be those of actual or potential comparative advantage for developing countries and for which regulations are important. The research team has chosen the following list of sectors, which cover several areas of regulatory scope.

Fresh produce: food safety, human health, SPS.

Processed foods: food safety, human health, SPS.

Electrical and electronic equipment: electrical safety.

Industrial or agricultural chemicals: environmental protection.

Textiles and apparel: environmental protection.

Telecommunications terminal equipment: electromagnetic interference.

Automotive parts: passenger safety.

TBTs Selection. In the main, the standards about which the surveys will generate information are dependent on respondents' answers. Specific firms may pay more attention to standards of particular concern to their product lines than to broader sectoral or national regulations. Yet for comparative purposes it is useful to develop information about the operation of identifiable regulations in both domestic markets and export markets. For example, it is interesting to discover whether the responses of South African and Egyptian exporters to a British regulation are different and whether those responses are further differentiated between the British and the American regulation. Conformity assessment procedures may be particularly important in this context. Thus, for each sector incorporated into the analysis it is desirable to characterize the major TBTs in important export markets. This will further help interviewers guide respondents' attention to international cost comparisons. Selected export markets will include the United States, Japan, Canada, Australia, the European Union (to the extent there are common regulations), or specific EU countries, such as Germany, the U.K., and France.

Time Selection. It will be important to repeat the surveys in at least two respondent countries annually over a three-to-five year period. With a sufficiently large number of firms in the surveys the panel results could be used to assess changes over time in responses to standards and export behaviors.

Question Categories. The survey instruments should elicit the following types of information from each firm:

1. Production costs, including fixed costs. Fixed costs are charges related to headquarters and plant operation that are unaffected by output scale. Examples include R&D expenditures, marketing costs, debt service, property taxes, and certain capital taxes. Variable costs are charges for inventory, materials, labor, new investment, and taxes related to profits, outputs and variable inputs. Estimates of installed capital values are useful for identifying short-run fixed factors.
2. Output measures, including domestic sales, exports, and exports to particular markets in which TBTs are to be studied. The dates at which firms entered production and export markets are important, as is information on why firms choose particular export markets or choose not to export.
3. Measures relevant for assessing market structure, including numbers of firms by sector and number of importing firms; firm sizes by employment, capital stock, and sales volumes; and number and age of product lines.
4. Major impediments to firms seeking to expand output, exports, or enter new markets. These impediments may be unrelated to policies. Examples include distance to markets, scarcity of skilled labor, translation costs, and fixed exporting costs. They also may stem from policies, including restraints on hiring and firing, inadequate access to capital due to financial market regulation, protected and inefficient transport systems, export disincentives, and trade restraints in import markets.
5. Assessments by firm managers of the costs of compliance with TBTs. Questions could be posed in a multidimensional framework, as shown in Table 2.2. First, for each regulatory area (regulations in home country, importer regulations, conformity assessment, and MRAs) the standards or regulations at issue must be precisely identified. In most cases the respondent would identify the procedures of most concern to him. However, in the case of importer regulations, conformity assessment, and MRAs, it would be important for the interviewer to ascertain whether the firm must deal with those procedures that were identified for key export markets in the standards selection stage. Second, there are issues of consistency and whether the rule induces higher quality. Finally, to the extent possible it would be valuable for the survey to ask about how firms respond on the cost side, including whether production processes vary by intended market, impacts on input use, and effects on fixed costs.

Supplemental Information

To facilitate the empirical analysis the survey results could be supplemented by other sources of information. First, detailed information on standards and technical regulations is available from international sources.

A prominent case is PERINORM. However, the country coverage of this database is limited and it does not incorporate SPS measures. Thus, for the target countries and sectors it will be necessary to develop further compilations of TBTs, SPS procedures, and standards.

Next, to undertake analysis of TBTs and trade at the detailed sectoral level we need bilateral trade data for the countries in the standards and regulation databases. Trade data are easily available but, since they are classified on the Harmonized System, they must be concorded to the sectoral classification of standards. The trade data would also be useful for constructing weighted averages of the standards counts data in an effort to assess the relative importance of standards by type.

Finally, it is of considerable interest to supplement the survey data on outputs and costs with information from the statistical authorities of target nations on detailed sectoral production and costs by country.

TABLE 2.2. Basic Design Survey Questions Regarding TBTs

Home Regulations	Importer Regulation	Importer Conformity Assessment	MRAs or Harmonization
Identification	Identification of countries and regulations	Describe requirements and operation	Identification of agreements
National or regional and consistency	Consistency across markets	Consistency across markets	Consistency across MRAs
Impact on quality	Impact on quality	Investments in quality control	Preferences for MRAs or harmonization agreements
Costs of re-design	Costs of re-design	Costs of compliance	Impacts on costs
Home sales vs. exports	Differences by export market	Differences by export market	Differences by MRA
Input use	Input use	Internal vs. external resources	Compliance costs
Investment and R&D	Investment and R&D		Exclusionary impacts

This would provide another layer of information for investigating how regulations affect productivity, product quality, and costs and also would permit comparison of the survey data for consistency with other sources. These data need to be assembled on a time-series basis for as many countries and sectors as possible for the analysis. Thus, one desirable criterion for country selection is that the country undertakes periodic or annual detailed industrial surveys. Such information might also be collected from international organizations, such as UNIDO.

Empirical Approaches

With the information described it will be possible to engage in substantive empirical inquiry. An initial overview of the data would be informative about such questions as how firms respond to standards and regulations, correlations between regulations and costs, and the incidence of regulations between home markets and export markets. However, our interest lies in deeper econometric and structural approaches.

Among the questions to be analyzed are the following. First, what are the effects of particular standards on productivity, costs, and export opportunities? Second, can these cost measures be used to develop meaningful indicators of trade restrictiveness? Third, are there significant diversionary effects from bilateral or regional MRAs and standardization blocs? Under what circumstances would this pose a danger?

Gravity Models of Trade and Standards

One approach is to extend the study by Moenius (1999), which was a simple gravity model of trade. In the standard gravity approach, bilateral trade between countries, either at the aggregate level or at the industry level, is regressed on the GNP levels of both countries, their populations, and geographical distance between them. Moenius did not incorporate these variables into his estimation, attempting to capture them, and other trade determinants, with a series of dummy variables. In that context his results are difficult to interpret for they may not adequately control for scale effects and resistance factors. A particularly noteworthy omission was the absence of indicators of joint membership in regional trade agreements. Thus, one important extension would be to include these, and other, basic trade determinants into the specification. Further, Moenius included only counts of product standards in his estimation, making it impossible to distinguish between standards and technical regulations. To the extent we can gather information on such regulations, including them in the trade equations should provide valuable information about differences between the trade effects of voluntary standards and TBTs. Finally, our aim is to incorporate trade and regulation activity in a number of developing countries.

Thus, a useful form of the gravity equation is:

$$\begin{aligned} ln(M_{ij}^{k}) = {} & b_0 + b_1 ln(GNP_i) + b_2 ln(POP_i) + b_3 ln(GNP_j) + b_4 ln(POP_j) + \\ & b_5 ln(DIST_{ij}) + b_6 RTA_{ij} + b_7 ln(ST_i^k) + b_8 ln(ST_j^k) + b_9 ln(ST_{ij}^{k}) + \\ & b_{10} ln(TB_i^k) + b_{11} ln(TB_j^k). \end{aligned} \qquad \textbf{(1)}$$

Here, subscripts i and j refer to exporting country and importing country, respectively, while superscript k indicates sector of activity. *RTA* denotes a dummy for bilateral membership in a regional trade agreement and is taken here as a prototype variable for trade-expansion factors. The variable *ST* refers to standards, which, as in Moenius (1999) may be exporter-specific, importer-specific, or shared between the nations. The variable *TB* refers to technical regulations on trade in manufactures and sanitary standards in foods. The conception here is that these standards and regulations may be identified at the sectoral level, though national measures might be used as well. Note that sector-specific impacts of standards on trade may be captured by individual sectoral regressions or by sectoral fixed effects in a pooled regression. The gravity equations would be estimated in a cross-section of countries and industries for a particular year.

The signs and significance of the coefficients should provide information on how standards and regulations affect trade volumes and how these impacts vary by sector. Of particular interest is whether standards and regulations differ. It is possible, for example, that standards could expand trade in certain circumstances, while regulations could limit trade. At a more quantitative level, the coefficients may be used to compute implied elasticities of trade volumes with respect to standards and regulations and whether such elasticities depend on the existence of preferential trade arrangements.

The primary measures of standards and regulations must be how many there are ("counts") by sector or country. It is generally infeasible to assess the relative importance of different standards simply by reading them. It may be possible to supplement the count data by developing qualitative measures of the extent and severity of standards from firm surveys. That is, firms could be asked to provide a numerical ranking by importing country and sector of how many standards must be accommodated and their intrusiveness. However, such indexes constructed from survey respondents should be treated with caution. For example, it is unlikely that firms would have sufficient information about restraints in countries to which they do not export to form meaningful opinions. Moreover, the subjectivity of such answers would depend importantly on other business conditions faced by firms.

Cost-Function or Production-Function Estimation

The essential question posed by this research is how technical regulations affect production and export costs. If the data collection is successful it will provide cost information for individual firms that may be used to estimate cost functions across firms within a sector or pooled across sectors. Assuming a time dimension exists for the data, a general specification of cost is:

$$C_{it} = G(Y_{it}, \boldsymbol{w}_t, \boldsymbol{Z}_{it}, t) + \pi_i \tag{2}$$

Here, C_{it} is total cost of production for firm i in period t, defined as the sum of primary input costs and intermediate purchases incurred to produce output and undertake necessary inspection procedures. Y_{it} is output, $\boldsymbol{w}_t$ is a vector of purchased input prices (including input taxes), and t is a time trend to capture technical change. The term π_i is a firm-specific productivity effect, estimable only in a pooled data set. It is reasonable to assume that common input prices face each firm. The vector $\boldsymbol{Z}_{it}$ captures additional impacts on costs, including such variables as public inputs (shared by all firms) and specification of fixed factors (for a short-run cost function).

A dual specification that encompasses the same technological information as the cost function is the production function with general representation:

$$Y_{it} = F(\boldsymbol{x}_{it}, \boldsymbol{Z}_{it}, t) + \theta_{it} \tag{3}$$

Here, $\boldsymbol{x}_{it}$ is a vector of inputs and θ_{it} is a different measure of firm-specific efficiency. As noted earlier, technical regulations could affect either fixed costs or marginal costs. If such distinctions may be drawn from the survey results, their separate impacts may be estimated via introduction of firm-level responses into a short-run cost function. Thus, two types of responses may be envisioned. First, for a given level of output firms may alter their demands for labor, capital, and materials in their marginal production costs, a result from, say, a regulation requiring inspection of all products and new packaging. Such impacts could be estimated by appropriate adjustment of the implicit factor-demand equations in the underlying cost and production functions. Factors devoted to compliance would be included as separate inputs.

Second, for a given production technique, firms may devote more expenditures to fixed costs, such as R&D to reorient production processes, factory inspection and maintenance, legal fees, and meeting reporting requirements.

These effects could, in principle, be estimated by altering the demands for fixed factors. Alternatively, to the extent that average costs are raised (average productivity is lowered), it should be feasible in the pooled approach to explain statistically the cross-firm or cross-sectoral productivity coefficients as a function of these cost shifts. Thus, π and θ could be specified as linear in the logarithms of standards costs, with a residual inherent productivity effect.

The cost or production functions selected for estimation would depend partially on the information provided by surveys and supplemental data sources. Given the multiplicity of cost factors, the most appropriate approach would be a flexible functional form, such as the translog or generalized Leontief (Kohli, 1991). However, because these functions are quite demanding in their informational requirements, it may be necessary to employ a more restricted approach, such as functions with constant elasticity of substitution.

Implementing the cost or production approach will be a challenge in the event that data are limited or essentially qualitative. For example, suppose we considered the firm-specific productivity effects to have components reflecting a standards impact and a TBT impact, where these impacts could vary by export market and between exporters and non-exporters. This specification would require the data to support estimation of large numbers of parameters that must satisfy certain regularity conditions. Moreover, it may be that survey respondents are capable only of providing standards counts or qualitative information, such as a ranking of cost impacts as "low", "medium" or "high". In that event, the estimation must shift to a more qualitative basis, with a selection of discrete independent variables.

Simulation Analysis

Data collection and cost estimation should provide valuable evidence on the extent of regulations and their perceived or actual cost impacts. However, econometric cost estimation techniques place considerable demands on data that may be imprecise and noisy. Moreover, they cannot readily account for the potential impacts under conditions of imperfect competition or other market distortions. Neither do they permit consideration of demand responses in market countries where products are differentiated and consumers may interpret adherence to standards as a legitimate signal of quality. In this regard, both partial equilibrium sectoral models in each country, and a multi-country computable general equilibrium model, could be specified and simulated to provide additional information on the competitive aspects of standards. Indeed, the survey results and econometric parameters would serve as key inputs into this set of empirical exercises.

Partial equilibrium models consider impacts of standards and regulations within a particular sector, accounting for market structure elements, such as numbers of firms and the size distribution of firms.[8] Standards and regulations may be modeled as variables that either increase costs, shift demand, or both. Specification of demand impacts is important, for there could be both price-related shifts in demand for given substitution parameters (say, between exports and home goods) and a change in substitution possibilities. On the cost side, regulations could alter either marginal costs or fixed costs, while the impact on profits would depend on the ease of entry and exit and the nature of competitive conjectures. An additional parameter relates to market segmentation.

For example, one extreme specification would permit perfect competition with free entry, constant returns to scale without consideration of fixed costs, and perfect homogeneity of goods. In this framework, straightforward analysis of an increase in marginal costs would capture quantity, price and incidence impacts of a regulation, subject to assumed elasticities. It would also be straightforward to compute impacts on trade diversion from regional standards.

The competitive model is inconsistent with observed product heterogeneity in international trade, while firms may demonstrate increasing returns to scale associated with fixed costs. These costs are likely to be raised by certain types of technical regulations. Thus, another canonical model would permit firms to produce imperfectly substitutable goods, subject to free entry and increasing returns. In such a model long-run profits would be invariant to standards but the number of firms and the scale of operations would be affected. Alternatively, the model would be calibrated to existing numbers of firms with varying degrees of entry possibilities. These models could be implemented either with a fixed degree of product substitution on the demand side or with substitution elasticities that would vary with the nature of standards and regulations.

To illustrate, consider the following simple model of Cournot competition where imports and domestic goods are heterogeneous but an endogenous number of symmetric firms produce a homogeneous "national" variety. Within a market called "home" and denoted by H, both varieties H and F are sold (a similar specification holds in the rest of the world). The inverse demand functions are:

$$p = a - bqn + dq^*n^* \tag{4}$$

$$p^* = a^* - bq^*n^* + dqn$$

(5)

where *a (a*)* indicates the intercept for the home (imported) variety, b indicates the common slope coefficient on own demands, *d* is the common parameter detailing cross-substitution in demand, *q (q*)* is output (imports) per firm, and *n (n*)* is the number of home (importing) firms. Cost functions per firm may be specified as:

$$C = F + G + (\boldsymbol{wx} + c)q$$

(6)

$$C^* = F^* + G^* + (\boldsymbol{w}^*\boldsymbol{x}^* + c^* + t)q^*$$

(7)

Here, *F (F*)* is home (foreign) fixed cost per firm, while *G (G*)* is the addition to fixed cost associated with a technical regulation in the importing country. The variables ***w*** and ***x*** *(**w*** and **x***)* are input prices and per-unit input choices. The variable *c (c*)* indicates the addition to marginal cost of the home firm (importing firm) from a regulation in the importer. Finally, *t* indicates the marginal cost of importing, associated with transport costs and tariffs.[9] Firms would maximize profits subject to the demand and cost constraints, generating predictions for equilibrium levels of output, imports, prices, and firm numbers.

The associated first-order conditions could be calibrated to actual data on initial firm numbers, imports, domestic sales, and costs, along with demand parameters. The calibrated model could then be used to simulate the impacts of varying *G* and *c (G* and c*)* to determine impacts of standards and regulations on competition and trade. Note further that the intercepts *a* and *a** and the cross-substitution parameter d could be varied to capture the demand-side effects of standards.[10]

This basic model could be altered in a variety of ways to study the impacts of regulations in alternative market settings. For example, the case just discussed presumes that the importer imposes its standard on the exporting firm but that this has differential cost impacts on the firms, or that the standards themselves are different. Standards could be modeled to have the same cost impacts (a natural definition of harmonization). Alternatively, a mutual recognition agreement could be modeled by having country-specific inspection costs subject to a minimum quality constraint (that is, a maximum value for d). A separate specification could examine cases where two countries form a standards union (such as an MRA) and exclude a third country.

Finally, turning to a CGE framework, its main advantage lies in the ability to assess the cross-sectoral impacts of regulations on outputs, prices, employment, and trade, along with meaningful computations of economic welfare. CGE models may also be developed to handle alternative market structures, demand specifications, and policy interventions in a flexible manner, both for single countries and multiple countries. For example, an integrated multi-country model could examine the implications of an MRA between the United States and Europe on the exports of particular developing countries. Moreover, conformity assessment costs could be modeled as additional production activities in the intersectoral and international context.

Its disadvantage lies in the need to conduct the analysis at aggregated levels, making it difficult to translate specific regulations, which typically exist at the product level, into meaningful policy experiments. The chapters in this volume by Ganslandt and Markusen and by Anderson and Nielsen illustrate this problem. However, the data gathered through this project's surveys should support a considerable analytical improvement in that the reported costs for particular regulations could be combined into weighted-average cost wedges per sector.

Thus, assume that the data analysis generates information on domestic and foreign regulations and their direct costs in key nations, and that these may be aggregated into sectoral cost impacts. Assigning these wedges to countries and regions in could support computation of their general-equilibrium international effects. It would further be possible to compute impacts of preferential agreements on standards and regulations. Note that care must be taken in considering the welfare gains and losses from such exercises, for a simple treatment of standards as costs means they would reduce economic well-being. Thus, the analysis would require some modification of utility functions to account for the demand for standards.

Concluding Remarks

One purpose of this volume is to suggest approaches for practical analyses of standards and trade that could benefit understanding of these processes on the part of developing countries. These suggestions would inform the World Bank's research program on standards and technical barriers to trade. Our reading of the literature points to three avenues for research.

First, there remains considerable room for further applied theoretical work aimed at improving our understanding of how particular aspects of standards formation and operation affect behavior. The available literature tends to focus more on strategic aspects of international competition subject to cost-increasing standards issued by one country or a bloc of countries and less on the actual competitive impacts of particular standards. It would be fruitful, for example, to contrast the potential costs and benefits for lagging nations of harmonization versus mutual recognition agreements. Such policies might be complementary or offsetting in their effects on trade. They might also serve to exclude poor countries from the gains from standardization among richer nations, but we should know more about the circumstances under which that would be true and prospective policy responses.

Second, evidence on how firms in developing and transition economies actually respond to domestic and international standards is completely unavailable. Aggregate data on variables such as standard counts are useful for looking at broad trends but it is vital to undertake detailed microeconomic, firm-level studies of compliance costs, problems in compliance, quality responses, and export and import impacts. This would require extensive surveys of firm behavior in a cross-section of industries, with parallel studies in a sample of countries. The selection of countries and firms would be constrained by available research funds. However, a useful approach should incorporate countries at different levels of development and global integration, which presumably would correlate with the number and extent of domestic standards and the use of international standards. It should include also a sample of narrowly defined industries of export interest to the countries selected, such as certain agricultural products, apparel, and electronic implements.

Thus, the initial task will be to design and implement the recommended survey instruments. Besides country and industry selection, it will be critical to develop a questionnaire/interview format that bears promise of extracting important quantitative information that could be combined with other economic data in ensuing analysis. Regarding standards, questions to be posed should be aimed at determining the mix of domestic and international standards firms face, their compliance procedures and costs, related investments in quality and marketing, investments in re-design of products or production methods, and the like. It would be beneficial to undertake surveys over multiple years in multiple countries.

Finally, if these surveys yield usable and comparable information on costs, investments, trade, and growth they could form the basis for valuable econometric and simulation studies. Such studies could be aimed at a variety of questions. Are standards a determinant of trade expansion or contraction and under what circumstances? Is this process more or less in evidence for countries at varying levels of development? Which standards are critical in the determination of trade, structural change, and growth? Can the standards and measures of costs be combined in meaningful ways to develop some form of restrictiveness measures? A basic but valuable approach would be to calculate "tariff or subsidy equivalents" of particular standards in selected importing countries or regions and see if they vary by economic circumstances of trading partners. If such calculations could be made they would be informative in many contexts, including the potential effects of standardization blocs.

NOTES

1. The same situation applies to patents count data.
2. Indeed, the response is so great that it defies belief, given that 100 additional standards would represent only a small rise in the British standards count. It seems likely that the authors misinterpreted the units inherent in their data or estimation.
3. This distinction among standards is similar to conflicts in intellectual property protection between cost-raising market power and cost-reducing certainty benefits (Maskus, 2000b).
4. The Gasiorek, et al approach predicts a fall in intra-EU trade volumes after price integration because production for domestic markets avoids transport costs.
5. See Deardorff and Stern (1998) for further analysis.
6. See James and Anderson (1998).
7. Deardorff and Stern (1998) discuss this problem in the context of import quotas.
8. See Richardson (1993), Maskus (2000a), and the papers in Krugman and Smith (1994) for examples and Helpman and Krugman (1989) for a theoretical basis.
9. This model could be used also to investigate the determinants of FDI in the presence of regulations if we added plant-specific fixed costs.
10. While pre- and post-standard consumer surplus measures could be computed, they would not be particularly meaningful unless there were a fuller specification of utility in a general-equilibrium model.

REFERENCES

Deardorff, A.V. and R. M. Stern. 1998. *Measurement of Nontariff Barriers*. Ann Arbor: University of Michigan Press.

Gasiorek, M., A. Smith, and A. J. Venables. 1992. Trade and Welfare -- A General Equilibrium Model. In *Trade Flows and Trade Policy after 1992,* ed. L. A. Winters. Cambridge: Cambridge University Press.

Harrison, G.W., T.F. Rutherford, and D.G. Tarr. 1996. Increased Competition and Completion of the Market in the European Union: Static and Steady State Effects. *Journal of Economic Integration* 11:3; 332-365.

Helpman, E. and P. R. Krugman. 1989. *Trade Policy and Market Structure*. Cambridge: MIT Press.

Henson, S., R. Loader, A. Swinbank, M. Bredahl, and N. Lux. 2000. *Impact of Sanitary and Phytosanitary Measures on Developing Countries*, Center for Food Economics Research, University of Reading, Reading, UK.

James, S. and K. Anderson. 1998. On the Need for More Economic Assessment of Quarantine Policies. *The Australian Journal of Agricultural and Resource Economics* 42:4; 425-444.

Kohli, U. 1991. *Technology, Duality, and Foreign Trade*, Ann Arbor: University of Michigan Press.

Krugman, P. and A. Smith, ed. 1994. *Empirical Studies of Strategic Trade Policy*. Chicago: University of Chicago Press.

Maskus, K.E. 2000a. Intellectual Property Rights in Lebanon. In *Sizing Up the Competition*, ed. B. Hoekman and J. Zarrouk. Ann Arbor: University of Michigan Press.

Maskus, K.E., *Intellectual Property Rights in the Global Economy*. 2000b. Institute for International Economics, forthcoming.

Moenius, J. 1999. Information versus Product Adaptation: the Role of Standards in Trade. manuscript.

Organization for Economic Development and Cooperation. 1999. *An Assessment of the Costs for International Trade in Meeting Regulatory Requirements*. TD/TC/WP(99)8/FINAL. Paris: OEDC.

Otsuki, T., J.S. Wilson, and M. Sewadeh. 2000. "Saving Two in a Billion: A Case Study to Quantify the Trade Effect of European Food Safety Standards on African Exports," Washington D.C.: World Bank.

Perroni, C. and R. Wigle. 1999. International Process Standards and North-South Trade. *Review of Development Economics* 3:1; 11-26.

Richardson, J.D. 1993. *Sizing Up U.S. Export Disincentives*. Washington DC: Institute for International Economics.

Swann, P., P. Temple, and M. Shurmer. 1996. Standards and Trade Performance: The UK Experience, *Economic Journal* 106: 1297-1313.

United States International Trade Commission. 1998. *Global Assessment of Standards Barriers to Trade in the Information Technology Industry*. Publication 3141, Washington, DC: USITC.

CHAPTER 3

Regulatory Protectionism, Developing Nations and a Two-Tier World Trade System

Richard E. Baldwin

Introduction and Motivation

Army generals, it is said, are always preparing to fight the last war they won. The WTO is in danger of following the same logic. The WTO/GATT has all but won the war against tariffs. Tariffs on OECD imports of manufactured goods – which account for two-thirds of world trade – average only 3.8%, with duty free treatment applying to fully two-fifths of these trade flows. As in all wars, victory is never total. Tariffs on rich nation food imports and poor nations industrial imports are still high, but these items account for little of world trade. Moreover, at least the poor-nation industrial tariffs seem to be falling of their own accord. Trade, nonetheless, is far from unfettered.

'Regulatory protection' is one name for the tens of thousands of cost-raising, behind-the-border measures that continue to substantially inhibit trade. Most of these measures are seemingly innocuous when viewed individually, but tangled together they are able to significantly fragment world markets. This point is not new. The last time 'globalization' was in fashion – when it was called internationalization and interdependence – Baldwin (1970 p.2) wrote: "The lowering of tariffs has, in effect, been like draining a swamp. The lower water level has revealed all the snags and stumps of non-tariff barriers that still have to be cleared away." The intervening thirty years have witnessed completion of the swamp-draining, but the stumps have started to grow; three decades of every tighter regulation of goods – most of which was adopt for purely domestic policy aims – have escalated regulatory protection.

Trade nevertheless is, overall, much freer than it was when the GATT was signed in 1947. Yet paradoxically, this means that remaining barriers have a more important, not less important, impact on production. The magnification effect of globalization" is the name for this. Falling tariff levels, teamed with lower transportation and communication costs, have cleared away many manmade and natural obstructions to trade, making the international 'playing field' flatter than it has ever been.[1] This very flatness, however, means that even a slight 'tilt' tends to have large effects on the location of production. In particular, seemingly minor differences in technical norms can have an outsized effect on production. While it is easy to exaggerate how footloose is international production, one thing is clear. As the field gets flatter, regulatory protection's influence on production will continue to grow, forcing its liberalization ever higher on the negotiators' agendas.

In this chapter, I analyze certain aspects of technical barriers to trade (TBTs), which is the phrase used to describe regulations and standards, imposed on product characteristics or production methods, that may impede international trade. According to common usage, regulations are mandatory while standards are voluntary.

The main thesis of this chapter is built in five steps. First, TBTs are important. Second, TBT liberalization will continue. Third, this liberalization will involve "hegemonic harmonization" or mutual recognition of rules and test. Fourth, such liberalization will almost surely entail preferential arrangements among rich nations, creating in essence a two-tier system of market access with developing nations in the second tier. The final step is to conclude that the WTO should be modified to address the potentially discriminatory aspects of regional TBT liberalization initiatives. The WTO, in short, should be gearing up to "fight" the battle of frictional barrier liberalization.

The remainder of this chapter is in six parts. Section 2 defines and describes TBTs and covers the basic economics of their impacts, focusing mainly on the discriminatory aspects of plurilateral liberalization. The section also reviews available empirical evidence. Section 3 discusses various TBT liberalization experiences, since a thorough understanding of the historical lessons is crucial to predicting the path of future liberalization. Section 4 argues that the main difference between TBT liberalization and tariff liberalization is based on obscurity rather than sovereignty. This basic point is important in understanding why TBT liberalization based on multilateral negotiations is almost sure to fail. The section also presents a political economy model that identifies large corporations engaged in two-way trade in similar products as the drivers of TBT liberalization. Section 5 draws the conclusion that continued regional TBT liberalization may create a two-tier structure of market access with rich nations systematically enjoying better access to each other's markets. The section also makes a number of proposals for WTO reforms that could help redress this problem. The final section presents a summary of the chapter's findings.

TBT Liberalization is Important

What are TBTs?

Technical barriers to trade result from norms (regulations and standards) that control the sale of goods in a particular market by specifying required product characteristics or production processes. There are two distinct aspects of this control: (i) content of the norm, and (ii) testing procedures necessary to demonstrate that a product complies with the norm. Thus, they come in two basic forms, content-of-norm TBTs and testing TBTs. The mind-numbing minutia and technical complexity behind most regulations and standards makes it difficult to speak of TBTs as a whole. Some examples may help. We start with a content-of-norm example.

All cars sold in Sweden must have wipers on the headlights. This policy may have started as a meritorious safety regulation (Swedish motorists keep their lights on at all times for safety. In the old days Sweden had lots of dusty roads). Today it mainly serves to raise the relative price of imported cars. From the drawing board onwards, Volvos and Saabs – and their factories – are designed with headlight wipers in mind. For other carmakers, Renault for example, the Swedish market is too small to really matter, so Renaults are not optimized for headlight wiper installation. Thus while it is expensive to put headlight wipers on both Swedish and French cars, the regulation raises the relative cost of French cars in Sweden. This gives Swedish carmakers an edge in Sweden. Indeed, the outcome may well be that Renault will only export to Sweden luxury models for which headlight wipers are already an option, so the regulation may act as a zero import quota for low-end cars.

Testing can also lead to TBTs. A frequent source of US-Japan trade tensions in the 1990s was asymmetric testing procedures. For example, Japanese manufacturers, who were trusted by Japanese regulators, could establish compliance with regulations by having a few units tested periodically while foreign firms had to have each import shipment tested (Sykes 1995 p.25).

Vertical and Horizontal TBTs

It is useful to classify content-of-norm TBTs into two sub-types, horizontal and vertical. Vertical-norm TBTs involve norms that can readily be characterized by level of stringency. For example, the United States adopted a law (the so-called CAFE standard, for corporate average fuel economy) requiring each car company's fleet to attain certain maximum levels of emissions when averaged over all models sold within its borders. This norm would have forced European carmakers to invest in more emission-abatement than U.S. rivals, which lower their average, in part, by selling small-engine cars while

European car companies sell mainly high-end cars with big motors. Likewise, Europe has much stricter rules than the United States on the permissible level of growth hormones in beef. TBTs stemming from such 'vertical norms' generate much emotion and media attention since they can be portrayed as protecting local consumers from low-quality imports.

Horizontal norm differences, however, are probably more common. Many TBTs arise when a nation, or sub-national government, adopts the specifications of the local firm's differentiated product as its norm. For example, France and Germany regulate the nature of permissible electric plugs (both have two round pegs, but the German pegs are wider). The two plugs are equally safe and equally expensive to produce, but they are different. This raises the French firm's cost of selling in Germany (since German firms do not have to modify their product to meet the regulations while French firms do). It also raises the cost of German firms selling in France, so TBTs based on horizontal norm differences tend to create reciprocal trade barriers.

It is easy to see why horizontal TBTs arise so often. Citizen concern and industrial efficiency demand product norms and a typical rich nation will have tens of thousands of standards and regulations. Most TBTs are highly technical and a large fraction covers intermediate inputs, which are products unknown to most voters. Due to their technical complexity and political invisibility, product norms are often written, directly or indirectly, by domestic firms to which they apply. Naturally these firms write the norms in a way that favors their varieties or at least disfavors foreign varieties. The result is a horizontal content-of-norm TBT.

The mere fact that standards and regulations inhibit trade and competition is not an argument for their removal. Good governance requires regulation to protect the health, safety and well being of citizens, animals and plants as well as to facilitate market transactions. The main problem with TBT liberalization is that it is difficult to know whether a particular norm serves the public interests or protectionists' interests. Indeed, both motives are often combined in a single TBT.

Modeling TBTs

As the examples show, product norms and testing procedures can distort trade when they raise foreign firms' costs relative to those of domestic firms. In certain industries the impact of TBTs is radically more complex. In industries with network externalities, like mobile telephones, standards can be manipulated to throw up barriers against non-local firms. In industries with patent races, such as pharmaceuticals, a regulation that merely delays the introduction of foreign goods can significantly alter the market outcome in favor of home firms. In industries with learning curves, product standards that apply to only a fraction of the market – government or military purchases, for example – can have large effects on the market equilibrium. Modeling TBTs in such "sexy" industries undoubtedly will be the focus of much future work. However, in this chapter, I focus on mundane industries where TBTs act by raising the costs of foreign firms more than the costs of local firms.

The cost-raising aspects of TBTs can affect foreign firms' fixed costs of selling to the regulated market as well as their marginal costs of doing so. Figure 3.1 shows how a typical TBT involves both up-front, one-time costs, such as learning about the regulation and bringing the product into conformity, as well as ongoing costs, such as periodic testing, or higher marginal production costs that stem from a low scale of production.

The Economics of Regulatory Protection

There has been surprisingly little work done on the economic effects of TBTs, even at an abstract level. Most trade economists working in the subject tend to assume that TBTs affect trade in the same way as frictional barriers (i.e. tariff-like measures that generate no revenue or rents). Most of the simulation studies of the EU's 1992 program, for example, modeled the liberalization effects as lowering the real cost of intra-EU trade by about 2%.

FIGURE 3.1. Compliance costs profile over time

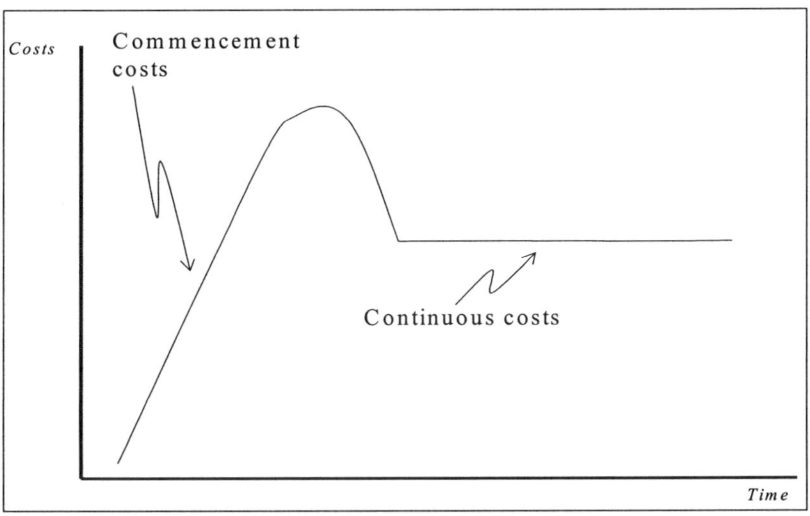

Source: Henson (1998) p. 18

Industrial organization literature has focused on the impact of standards on a narrow range of industries, namely industries marked by network externalities such as mobile phones, consumer electronics, computers and the like. As the survey by Matutes and Regibeau (1996) argues, the general presumption is that harmonization of standards within a single market can be manipulated by incumbents to raise barriers to entry. In such industries, regional TBT liberalization – which usually involves some form of harmonization of standards – would be presumed to act as a trade barrier to outsiders. While these network industries tend to attract a lot of attention, they cover a fraction of the trade affected by TBTs.

A Baseline Model

TBTs can affect fixed costs as well as variable costs, so even a minimalist model requires imperfect competition and increasing returns. With these elements, however, almost anything can happen theoretically. As Helpman and Krugman (1989) show, quotas can lower prices, export subsidies can improve welfare, and small countries can have positive optimal tariffs. Clearly then, we should not be surprised that, theoretically, TBTs have an ambiguous impact on trade flows and welfare.

To highlight the main sources of ambiguity, I lay out a baseline model. Consider an industry marked by Cournot oligopolistic competition with firms facing constant marginal costs and two types of fixed costs, a firm-specific fixed cost of setting up production (unrelated to TBTs) and a fixed cost of selling to each market (related to TBTs). To allow discrimination, we need three nations – home (H), partner (P) and rest of world (R), assumed *ex ante* identical to reduce complexity. The inverse demand function of a typical nation is $p_i=1-Q_i$, where p_i and Q_i are the price and total sales in market i (i=H, P,or R). The total quantity in market j is $(\Sigma n_i q_{ij})$ where q_{ij} is sales of a typical i-based firm in market j and n_i is the corresponding number of i-based firms. Marginal production costs are normalized to zero. The three markets are assumed to be segmented. Here we focus on TBTs based on horizontal norm differences and model them as simply as possible. Non-local firms face higher variable costs and/or fixed costs when selling to the local market. Firms located in home (H-based firms) face no additional costs when selling to the home market. Partner-based firms face a specific cost $0\leq\tau$ when selling to the H-market. The per-unit cost facing rest-of-world-based firms in the home market, τ^*, is initially identical to τ. The home and partner nations are always identical. TBTs may also present non-local firms with additional fixed cost. To capture this, we assume that each market has its own norm and complying with these costs F in each market (F is the same in each market since we are addressing horizontal norm differences).

The equilibrium is found by solving the nine segmented market first order conditions for the nine levels of sales. The model's linearity supports solutions for prices, consumption, welfare, number of firms, trade flows, and so on. In particular, the equilibrium price in market H, p_h, equals the sum of two terms. The first term, $1/(1+\Sigma n_i)$, reflects the level of overall competition. The second term, $(\Sigma n_i c_{ih})/(1+\Sigma n_i)$, reflects the average marginal cost of firms active in the market (c_{ih} is the market-specific marginal cost, normalized to zero for H-firms, to τ for P-firms and to τ^* for R-firms). The focus of the inquiry, however, is on the distortionary impact of TBTs, so I highlight the 'trade bias', defined as the difference between home's imports from all P-based firms and home's imports from all R-based firm. This bias roughly corresponds to a preference dummy in a gravity model equation.

When TBTs among all nations are symmetric (i.e. there is no discriminatory dimension favoring P-based firms over R-based firms), the bias equals zero. Specifically,

$$Trade\ \ Bias\ = p_h(n - n^*) - (n\tau - n^*\tau^*)$$

where n is the number of home and partner firms (always identical in the exercises we consider) and n* is the number of rest-of-world firms. In some cases, we allow the number of firms to vary with TBTs. The equilibrium expressions for n and n* are messy, but it is easy to show that the equilibrium number of home and partner firms falls with F and rises with F* while the equilibrium number of R-based firms falls as F* rises or F falls.

TBT Liberalization

TBT liberalization can take the form of lower fixed costs and/or lower marginal costs of selling to a particular market. Each of these may be "open" or "exclusionary". The trade impact of these various forms differs radically, as we shall see.

Open versus Exclusionary Mutual recognition:
Liberalization of Fixed Cost TBTs

We start by studying the liberalization of fixed-cost TBTs and, for simplicity's sake, we neutralize variable cost TBT discrimination by assuming $\tau=\tau^*$. Initially we suppose that the same fixed-cost TBT applies on an MFN basis, so R-firms, H-firms and P-firms have to pay a total of 2F to sell in both H and P. The liberalization considered is meant to resemble one aspect of the EU members' mutual recognition of each other's norms and testing. After the liberalization a product that complies with either home's or partner's norms, can be sold freely in both markets. That is, H-firms and P-firms have to pay only a total of F to access both the H and P markets. This mutual-recognition privilege may or may not extend to R-firms. On the one hand, mutual recognition can be "open", meaning that R-firms can also sell to both H and P markets after certifying their product in either. In this case, the fixed-cost liberalization benefits all firms equally because after liberalization the total fixed costs of selling to the three markets is reduced from 3F to 2F. This raises profits and attracts new entrants in all nations. Given symmetry, n and n* rise equally so the trade bias remains at zero ($\tau=\tau^*$ and $n=n^*$). In short, open fixed-cost TBT liberalization is not discriminatory, even when it is regional.

On the other hand, mutual recognition can be "exclusionary" in the sense that it applies only to products made in H or in P. In practice this exclusion is enforced by rules of origin. For example, the EU-Swiss Bilateral Agreements encompass TBT liberalization, but only Swiss-made goods circulate freely in the EU market after having been certified in Switzerland. In our model this means that H and P firms pay only F to access the combined H-P market, but R-firms must continue to pay 2F. This results in improved profitability for H and P firms, which will lead to a rise in their number. There is also a drop in the number of R-firms as the new H-based and P-based firms crowd some of them out. This creates a positive trade bias as can be seen by inspection of (1) combined with the facts that $\tau=\tau^*$ and $n>n^*$. In short, regional liberalization of fixed-cost TBTs is not discriminatory when it is "open", but it is discriminatory when combined with rules of origin.

TBT Liberalization that Favors Rest-of-World Firms

Open fixed-cost TBT liberalization can actually favor rest of world firms. Due to fixed market entry costs, not all firms will sell to all markets. Rather firms will sell to markets where their cost advantage makes it worthwhile sinking F. For example, an East Asian manufacturer of electronic components might not find it worthwhile to export to Europe if each European nation had its own product norm (each involving a compliance cost), but might find Europe very interesting with an EU-wide norm. In the context of our model, suppose that home and partner are geographically close and this situation would give H and P firms a distance-related marginal cost advantage (compared to R-firms) in the H and P markets. If this advantage is large enough, R-based firms may choose not to sell in either the home or the partner markets, since the operating profit they would earn on such sales would not cover the per-market fixed cost F. An open, fixed-cost liberalization, which lowered the fixed cost from F in each market to F for both the H and P markets could induce new R-firms to start selling in the H and P markets. The associated rise in n* (number of R-firms selling in the H and P markets) would <u>reduce</u> the trade bias.

To make the point more formally, start with the non-discriminatory situation where each firm must incur a fixed cost F to sell in each of the three markets. To simplify, suppose that TBT-related marginal costs are zero ($\tau=\tau^*=0$). However, assume that H and P firms have a natural advantage since R-firms face a trade cost of T when selling to the H or P markets, but H and P firms face zero trade costs when selling into each others' markets. From the first order conditions, a typical R-firm's sale to the H-market equals p_h-T, where p_h is the equilibrium H-market price. Sales of a typical P-firm in H will be higher, namely p_h. Because with linear demand operating profit is the volume of sales squared, R-firms may not find it profitable to sell to either the H or the P market when T is sufficiently large. Specifically, there can be a situation where $(p_h-T)^2<F$, so R-firms sell to neither the H nor the P market, but $(p_h)^2<F$, so H-firms and P-firms will sell to both markets. Now, consider the impact of an open, fixed-cost TBT equalization between the H and P nations. Post-liberalization, R-firms will sell to both H and P markets as long as the sum of operating profit earned in the H and P markets exceeds F. With symmetry, $2(p_h-T)^2>F$ is this condition, where p_h is the post-liberalization price. The impact of this case on the trade bias is straightforward. The liberalization has no impact on P-firm sales to home (since the liberalization only affected F and all P-firms were already selling to home), but it induces a whole new set of R-based firms to export to home. The trade bias, which was positive before the liberalization, now falls.

Liberalization of Variable-Cost TBTs

Regional TBT liberalization can easily affect the variable cost of complying with product and process norms. For some products, such harmonization may create a preference for EU products. The EU could, for example, adopt a common norm that is systematically more costly for non-EU firms to comply with due to scale economies or specific technological advantages among EU firms. This sort of case may arise when, for example, EU firms sell the majority of their output in Europe and so design their products and factories with the EU norm in mind. If scale economies are important, the per-unit cost of meeting the EU norm may be lower for EU firms than for non-EU firms for whom Europe is a minor market. Alternatively, adoption of a common norm and testing procedure might result in scale economies that would lower compliance costs for all firms.

In the formal analysis, the first type of harmonization – call it prejudiced harmonization – shows up as a change from the initial situation of $\tau=\tau^*$ to a situation where τ^* exceeds τ. This sort of regional liberalization has the usual 'supply switching' effect of inducing H and P consumers to switch away from R-made products and towards H-made and P-made goods. In particular, taking n=n*, for simplicity, the trade bias, which reduces to $n(\tau^*-\tau)$ in this case, is positive. In the second type of harmonization – call it neutral harmonization – $\tau=\tau^*$ both before and after liberalization, but the level of costs fall. Plainly, this will have no effect on the trade bias.

Trade-Reducing Regional TBT Liberalization

TBT liberalization may affect trade bias by altering firms' ability to price-discriminate internationally, for example by switching the nature of competition from segmented markets to integrated markets. As literature on trade policy and imperfect competition has made clear, the impact of protection can be radically different with and without international price discrimination (Markusen and Venables 1988). Only one such implication is examined here.[2]

Markets are segmented when firms can assume that their customers will not re-sell across markets. Markets are integrated when firms treat the two markets as one. Even when the two markets are symmetric and equilibrium prices are identical, market segmentation matters. In particular, the volume of trade is higher under market segmentation due to 'reciprocal dumping' trade. The relevant link with TBTs is that product regulations may help firms segment markets. Consequently, the imposition or liberalization of TBTs may have important effects that are not captured by standard NTB analysis.

To make the point as cleanly as possible, consider two identical nations, each with a single firm. The firms produce an identical consumer good at identical costs and trade in the good is subject to some natural transport costs. Initially, the two nations impose a safety regulation of a special type. For safety's sake, the good and its instructions must be sealed in tamper-resistant packaging. The instructions must be written in the language of the country, and only that language, to avoid confusing consumers. Presume that instructions would be included in any case and translation costs are minimal, so the regulation would have a negligible impact on marginal and fixed costs.

Using the analogy of tariff analysis, this regulation would seem to have no trade effects. This is incorrect. The regulation prevents re-sale by consumers or middlemen and thus allows each firm to treat the two markets as segmented. When the two markets are segmented, yet there are some natural trade costs, the two firms engage in reciprocal-dumping trade à la Brander and Krugman (1983). In particular, firms charge a lower producer price for their exports than they do for their local sales. If the regulation is changed to require, for example, instructions to be printed in both languages, the markets become integrated in the sense that re-sale across markets is possible. Such arbitrage would continue until producer prices are equalized.

Interestingly, this market integration has a dramatic and counterintuitive trade effect. All trade halts since each firm supplies only its local market. The reason is that, due to arbitrage, firms treat a sale in either market as if it were a sale in the local market. In this case there is no return to incurring the trade costs of selling abroad. Nevertheless, prices fall since the potential competition due to arbitrage is a more powerful source of competition than are actual imports without arbitrage possibilities.

Empirical Evidence

The economics profession does not yet have direct empirical measurements of the trade-inhibiting effects of TBTs in general.[3] There is hope that this lacuna will be filled by the World Bank project on this topic. At this point, indirect evidence is all that can be marshalled.

Sapir (1997) tried to gauge the impact of EU integration on non-member nations since 1960. To this end, he estimated year-by-year gravity equations for the 1960-1992 period on aggregate, bilateral trade flows among a set of West European nations. The key interest lies in the sign and size of the dummies he introduced for all the European preferential trade arrangements (PTA). Interpreting the dummy coefficients, however, requires some care. Because Sapir did not include a dummy for trade flows among the original 6 members of the EU, this level of integration was taken as the standard against which all other PTAs are judged. Thus, the signs of the included PTA dummies tell us whether the particular bilateral trade flow is higher or lower than it would have been if the two nations had joined the EU in 1958. Standard controlling variables, such as GDP of sending and receiving nation, distance and common language, were included.

It is important to note that all tariffs and quotas on intra-EU trade and all industrial-goods tariffs on intra-EFTA trade had been eliminated by 1968. All tariffs on industrial trade between the European Free Trade Area (EFTA) and the European Union were removed by the 1973 EU-EFTA free-trade agreements (industrial goods account for the vast majority of EU-EFTA trade). As a result, all of Western Europe was part of an industrial duty-free zone by 1973. The importance of this fact should be clear. Almost all of the West European integration since 1973 has involved the removal of frictional trade barriers, with TBT liberalization being an important component of this, especially during the Single Market Program (initiated in 1987). It is also important to note that the EFTA nations did not participate in the Single Market integration until the agreement establishing the European Economic Area (EEA) was finalized in 1993.

It would be interesting to measure directly the impact that this frictional-barrier liberalization had on intra-EU trade, but Sapir's results do not permit this. We can, however, garner some evidence on the trade-volume impact of the liberalization by looking at changes in the coefficient on the EFTA-EFTA and EC-EFTA dummies.

The annual point estimates are plotted in Figure 3.2.[4] According to the estimates, up to 1975 EFTA-EFTA trade volumes were higher than the EU-EU flows, controlling for the usual factors.[5] This somewhat unexpected result (the received wisdom is that the EU membership provides the tightest integration in Europe) may be due to Sapir's omission of an adjacency variable.[6] In any case, the point estimates on the EFTA-EFTA dummies are never significant (at the 95% confidence level) when they are positive. The EFTA-EFTA dummy coefficient becomes solidly negative in the early to mid 1980s, but become significantly less than zero only in 1989.

The rapid decline in the EFTA-EFTA coefficient, together with the fact that EFTA was not pursuing any significant intra-member integration initiatives, suggests that EU-EU trade was promoted by the Single Market's removal of frictional barriers. (Recall that the dummies indicate market access relative to the market access available on trade among the original six EU members). A second piece of evidence comes from the evolution of the coefficients on

FIGURE 3.2. Sapir's evidence

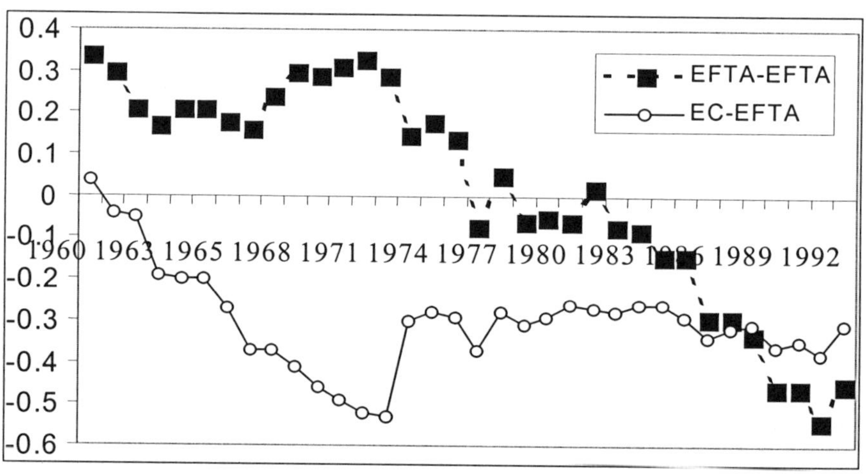

Note: zero indicated access at intra EU level
Source: Sapir (1997)

the EC-EFTA dummy. The jump in market access due to the 1973 agreements is clear in the point estimates. Somewhat less evident is the gradual decline of point estimates after the mid-1980s. The modest change in the point estimates, however, hides the fact that they become significantly negative in 1985 and remain so to the end of the period.

Head and Mayer (1999) found more conflicting evidence using industry-level data and the sophisticated gravity model approach inspired by Wei (1996). Head and Mayer used fairly disaggregated industry-level data (3-digit NACE) to estimate so-called border effects on purchasing patterns. In particular, the authors set out to answer three questions. How big were the border effects in Europe before the non-tariff liberalization embodied in the Single European Act? How closely were the industry-by-industry border effects associated with indirect measures of the severity of trade barriers? Has the change in border effects been associated with the removal of trade barriers under the Single European Act?

In addressing the first question, they pooled the 98 industries in their sample over the 1984-1986 period. Depending upon the exact specification, they found that individual EU nations buy between 13 and 20 times more from local producers than from producers based in other EU nations. This baseline regression provided no direct evidence on the cause of this border effect, but two facts are suggestive. First, note that there were no tariffs or quotas on intra-EU trade during this period, so non-tariff barriers are a likely culprit. This suspicion is reinforced by the fact that the regression results also showed EU membership stimulated trade by approximately 50%. The point is that the only non-member nations in their sample, Spain and Portugal, had enjoyed duty-free industrial trade with the EU since the 1970s, yet they did not participate in the EU's factor-market and non-tariff barrier liberalization. One possibility is that the impact of EU membership reflects the trade-inhibiting effects of the EU's preferential liberalization of non-tariff barriers and the border effect reflects the impact of remaining non-tariff barriers.

The authors, however, did not find any correlation between industry-specific border effects and two measures of intra-EU trade barriers. The first measure was based on a 1980s survey of EU firms and the second on an educated guess by Buiges, et al (1990). This absence of a correlation presents a puzzle. Industry-specific border effects are not correlated with proxies for trade barriers yet EU membership has an important impact on trade flows. This suggests one of three interpretations. First, deep EU integration could matter but the trade-barrier proxies are faulty. Second, deep EU integration could matter but its influence is due to something other than TBTs, though similar to non-tariff barriers. Third, NTBs do not matter and the Spanish and Portuguese trade flows are unusually low for some reason that is unrelated to trade barriers. Unfortunately, the authors' results do not allow us to distinguish among these inferences.

Results pertaining to the third Head-Mayer question are also interesting. Year-by-year regressions with pooled data showed a marked decline in the border effect, suggesting that policy changes or transport cost reductions have been important. However, the decline seemed to be much sharper before the Single European Act came into force. Again the interpretation of this finding is ambiguous. Either the Single European Act was ineffective at liberalizing regulatory protection, or it was effective at removing the barriers but the barriers had little impact on trade.

One of the most direct attempts to measure the trade impact of TBTs is by Moenius (1999). The paper focused on the trade impact of standards (voluntary norms) rather than on TBTs more generally due to data limitations. Because this paper and its results were discussed in Chapter 2 of this volume, I do not consider it further here.

Finally, it is worth citing some important but highly indirect evidence. Governments and firms engaged in international trade act as if TBTs have important trade-inhibiting effects. They also act in a way that suggests they believe that preferential TBT liberalization has an important discriminatory trade effect. As the discussion of TBT liberalization in section 3 below makes clear, the TBT liberalization in the EU's Single European Act induced non-EU governments nations to negotiate new trade arrangements. Notable among these are EU membership, the European Economic Area agreement and the EU-Swiss Bilateral Accords, which redressed the discrimination arising from the Single Market Program. Moreover, the rapid spread of mutual recognition agreements between the EU and its major trading partners, such as the United States, Canada, and Australia shows that firms and governments take the trade-inhibiting impact of TBTs seriously enough to invest time and energy in lowering them.

Given this admittedly scant empirical evidence that TBTs are important, we turn next to the real-world TBT liberalization schemes that have been pursued in the past. The lessons from historical attempts are crucial to predicting the path of future TBT liberalization.

Liberalization Experiences

TBTs inhibit trade by raising costs faced by foreign firms more than those faced by domestic firms. Liberalization requires a lowering of the cost gap and involves two main dimensions, content-of-norms and conformity-assessment. Liberalization of the first type involves making product norms more cosmopolitan and thus narrowing the cost advantage of domestic firms.

Liberalization of the second involves lowering the excess costs that foreign firms face in demonstrating compliance of their goods to accepted norms.

There are two ways forward, harmonization and mutual recognition. The first practice produces convergence to a single norm or conformity assessment procedure and the second generates acceptance of foreign norms and conformity procedures. Harmonization can be accomplished either via negotiation, through narrowing the differences between various nations' norms, or via the hegemonic route, where other nations adopt the norms of a hegemon.[7]

We turn first to the European experience for one simple reason. I argued earlier that the growing prominence of TBTs on the world trade agenda is due largely to the near-total victory over the 'easy' barriers, tariffs and quotas in OECD nations. Intra-European trade reached this stage more than three decades ago. Subsequent liberalization interest has focused on behind-the-border measures, with TBTs being one of the chief attention-getters. More generally, since trade integration among European nation is two or three decades ahead of WTO integration, Europe's experience with thorny issues, such as labor standards and trade, trade and competition policy, trade among rich and poor nations, and dispute settlement provides a natural departure point for the study of most 'new' trade issues.

EU Initiatives

The EU recognized TBTs as a barrier to the Common Market in 1957 with Article 100 of the Treaty of Rome.[8] This requires approximation (Eurospeak for harmonization) of national regulations for the "proper functioning of the common market". As in the WTO/GATT system, Europe's first liberalization efforts focused on the 'easy' barriers, tariffs and quotas. With these eliminated by 1968, liberalization attention turned to TBTs.[9]

Old Approach, 1969-1984

The EU systematically took up the removal of technical barriers in 1969 with its 'General Program', which launched what came to be called the 'old' approach to TBT liberalization. The old approach was, in essence, based on the idea that the EU would become as a unified economic area, functioning much like a single national economy. Specifically, the General Program comprised four Council resolutions and a framework decision adopted on 28 May 1969. The resolutions concerned a detailed timetable for a large number of directives on industrial products and foodstuffs, expression of the Council's intention to institute mutual recognition of conformity assessments, and a procedure for adapting directives to technological advances. The framework decision prescribed a standstill of Member State measures concerning products covered by the General Program and a requirement that Member States inform the Commission of new provisions for products not covered. These prescriptions were largely ignored.[10]

In short, the old approach dealt with the content-of-standards issue via negotiated harmonization. It strove for the adoption of a single standard laid out in detailed technical regulations for single products or groups of products implemented by unanimously agreed 'directives' of the European Council (the EU's main decision-making body).[11] Once adopted, such directives supplanted national laws, regulations and standards.

The conformity-assessment issue was tackled in a more decentralized fashion. For products on which directives were adopted, EU members recognized conformance assessments performed by designated bodies in any EU nation. To prevent new TBTs, the so-called framework decision provided for standstill and notification rules on new national standards and regulations.

Evaluation of the Old Approach

"This approach to technical harmonization failed completely, " according to Majone (1994), although it deserves credit as the world's first serious attempt to tackle TBTs. The detailed timetable was roundly ignored and harmonization of standards and regulations proceeded much more slowly than the development of new national TBTs. For example, ten years were required to adopt a directive on gas containers made of unalloyed steel and 9.5 years was the average delay for the 15 directives adopted *en masse* in 1984. Furthermore, rules on 'standstill' and notification were ignored since the Council qualified these rules as a 'gentleman's agreement'.

In practice, Member States had full discretion to adopt new national regulations, as long as they did not cross EU case law. The Commission received an average of only 11 notifications annually between 1975 and 1982 (Pelkmans 1989 p.109) while thousands of new regulations sprang up in Member States (Majone, 1994 p.166). The manifest failure of the old approach is confirmed by many experts, such as Michael Emerson (the chief Commission economist on the Cecchini Report) who stated that little progress was made since the exhaustively detailed directives were "difficult to agree and quick to become obsolete" (Emerson et al, 1989).

New Approach, 1985 to Present

The EU's 'New Approach', formally launched in January 1985, is something of a misnomer since it evolved over the 1960s and 1970s.[12] In particular, decisions of the European Commission and the EU Court of Justice introduced the main elements of the approach by the end of the 1970s.[13] In the 1974 "Procureur du Roi v. Dassonville" decision, the Court broadened the scope for Article 30 prohibitions. It opined that "all trading rules enacted by Member States which are capable of hindering, directly or indirectly, actually or potentially, intra-Community trade are to be considered as measures having an equivalent effect to quantitative restrictions," and are, therefore, to be prohibited. According to this ruling, virtually any regulation or standard that might hinder or substantially raise the cost of imports seemed to be prohibited. This decision did not automatically remove all TBTs. Each of the thousands of barriers would have to be challenged in court.

In the 1979 "Cassis de Dijon" case, the EU Court considered a German import prohibition on certain low-alcohol spirits. When challenged, the German government argued that the prohibition was necessary to protect public health (since weak spirits more easily promote alcohol-tolerance) and to protect consumers (since consumers might buy weak spirits, thinking they were strong). The Court ruled that the measure was not necessary since widespread availability of low-alcohol drinks in Germany made the prohibition ineffective in furthering public health (the Court thus dodged the issue of whether the German claim was disingenuous). It also found that putting the alcohol content on the label was sufficient to protect consumers, so the import ban was not "necessary".

In this case the Court explicitly allowed for some trade-inhibiting national regulations and standards. However, it set the permissibility test much higher for 'import-only' rules than for rules that applied national treatment to imports. For the former, Member States had to justify the provisions based on Article 36 exemptions (e.g. measures justified on grounds of public morality, public policy or public security, the protection of human, animal or plant health, protection of national treasures). The Court has narrowed interpretation of these exemptions to the point where only rules necessary to protect the health of people, animals or plants could plausibly be used to justify TBTs (Sykes 1995 p.98). For universally applied provisions, the ruling expanded the range of permissible, trade-inhibiting barriers. The main test for such measures was whether they were necessary to satisfy "mandatory" requirements, such as safeguarding the effectiveness of fiscal supervision, protection of public health, and defense of the consumer. The upshot of these rulings (and a knock-on Commission Directive interpreting the Cassis de Dijon case) was to make mutual recognition of other Member States' standards, regulations and conformity assessments the standard operating procedure. Any deviation would require clear and compelling justification and must pass the test of being the least trade-restrictive-means of accomplishing a "mandatory" requirement. The presumption in the EU is now that any product lawfully made or sold in any Member State can be sold in any other member state. Since 1994, the presumption also extends to Norway, Iceland and Liechtenstein under the European Economics Area agreement.

With this background, the basis of the New Approach is to distinguish sharply between the goals ('essential safety requirements' in EU jargon) and means ('harmonized standards') of product and process regulation. Two elements must be distinguished, 'mutual recognition' and 'technical harmonization'. For the vast majority of European products, TBTs are eliminated by the principle of 'mutual recognition' of testing and content-of-norms. EU case law and interpretations by the European Commission create the presumption that the standards and regulations of all Member States are merely different *means* of implementing equivalent regulatory *goals*. Thus, products made or sold in one Member State have 'single-market access'. That is, they can be sold freely and without further testing in all Member States (and in Norway, Iceland and Liechtenstein since the 1994 European Economic Area agreement). Member States are free to set standards and regulations on their territory, but any resulting barriers to trade are allowed only if they can be justified as being the least-trade-restrictive means of achieving a legitimate purpose, such as health, safety, consumer protection, or environmental protection. The EU Court has taken a very narrow view of allowable justifications.[14] Implicit in this approach is the mutual recognition of conformity-assessments performed by any body sanctioned by a member of the EEA.

For products where regulation-linked trade restrictions are justifiable, the EU has turned to technical harmonization. It is here that the new approach is truly new. In this framework, the goals are completely harmonized via Council directives that supplant national law. Detailed technical specifications and standards that show how to comply with these requirements are left to private European standard setting bodies, such as the Comité Européen de Normalisation (CEN) and Comité Européen de Normalisation Electrotechnique (CENELEC).[15] For example the Lifts Directive (95/19/EC), which is only 15 pages long, lays out essential health and safety requirements relating to the design and construction of lifts and safety components. Paragraph 1.2 in Annex I states "The car must be designed and constructed to offer the space and strength corresponding to the maximum number of persons and the rated load of the lift set by the installer." The Council leaves the listing of technical specifications that would fulfil this goal to CEN.

The effectiveness and political acceptability of this TBT liberalization is bolstered by two principles. First, the essential safety requirements must be transcribed into Member States' laws. Second, while the resulting harmonized standards must also be transposed at the national level – and all conflicting national standards must be withdrawn – the standards are voluntary. More specifically, goods made in accordance with harmonized standards are presumed to conform to the essential requirements and are thus free to circulate in the EEA. However, a firm may choose to deviate from these, say due to an innovation. In this case, single-market-access can be obtained by presentation of a certificate of conformity with essential requirements issued by a designated body (Pelkmans, 1989 p.105).

For products subject to technical harmonization (rather than mutual recognition), the conformity-assessment aspect of TBT liberalization is addressed by the 1989 'global approach'.[16] Four stages (design, production, marketing, and sales/service) are distinguished and at each stage conformity may be assessed by one of three bodies, the manufacturer, the local Member State's regulatory authority, or a designated third party certification body (Pelkmans 1989 p.107). Depending on the precise product in question, approval from the various standards-bodies must be obtained at various stages. Products for which this procedure is clear and for which a harmonizing 'essential requirements' directive exists can earn the "CE" mark. This mark ensures free access to all EEA markets.

The whole process of content-of-standards technical harmonization was expedited by the 1986 Single European Act (as implemented by the 1992 Single Market Program), which replaced unanimity with qualified majority voting in the European Council for all Internal Market decisions, including TBT issues. Majority voting also now applies to CEN and CENELEC decisions on the adoption of standards.

The new approach also made substantial progress on the prevention of new TBTs with the 1983 Mutual Information Directive. Member States must notify the Commission of new regulations. Both the Commission and other Member States can object to the regulation and request that it be modified to reduce its trade-distorting effects. In either case, adoption of the new regulation must be delayed and may be prevented altogether, if the Council adopts a harmonization directive in the meantime.

Additionally, national standard-setting organizations are required to notify the Commission and certain European standards bodies of proposed standards and regulations before their adoption. A standing committee can determine whether such standards would substantially interfere with trade and can then refer the matter to a European standards organization, or require the national agency to collaborate with interested Member States. When the matter is referred to a European standards body, Member States cannot enact their own standard until a European standard is set or until one year elapses. Some exceptions, concerning urgent health and safety considerations, are permitted. While this Directive is not radically different from the Gentleman's Agreement in the old approach, the Commission actively pursues cases of non-compliance because they are an infringement of Community law. A Council Directive has the force of a national law in all Member States and is enforceable by the EU Court. This notification, comment and objection procedure is credited with reducing the emergence of new TBTs in Europe.

Given the sparse empirical evidence available, it is too early to evaluate convincingly the New Approach. In terms of output, however, it has been a clear success. In 1975 there were 20 harmonized standards in the EU. In 1999 there were nearly 5,500 (Moenius 1999, p.1).

Despite this history, it is important to note that Europe's success with managed mutual recognition is not a good indicator of the route's success in the wider world. In particular, it would be fallacious to assert that such a system would be as politically acceptable elsewhere as it was in the EU after Casiss de Dijon. In the world at large, national firms still have considerable rents to gain by inducing their national regulators to adopt protective standards and regulations.

The Reaction of European 'Outsiders': EEA and Switzerland

TBT liberalization in the EU had discriminatory effects on the West European nations that were not EU members. In the late 1980s, EFTA firms and governments decided that they had to react. Several countries considered applying for EU membership (Austria actually did), while others considered bilateral negotiations. Jacques Delors forced the decision in January 1989 by proposing the European Economic Area agreement.[17] The agreement is complex and required the creation of extensive institutions, but its main importance was to bring the EFTA countries into the New Approach.[18]

While the effort to counter discrimination is easy to understand, the outcome is not. Two aspects of the EEA are extraordinary. First, it is unbalanced in terms of the rights and obligations of EFTA nations when it comes to future EU legislation. In essence, it forces the EFTA partners to accept future Council Directives (the *Acquis Communautaire*) concerning the Single Market, without formal participation in the formation of these new laws.[19] Second, the EU insisted on a good deal of supranationality among EFTA countries to simplify the task of keeping the Single Market homogeneous. The members of EFTA have resisted such supranational authority since the end of the second world war, so it is remarkable that they accepted it. In the end, virtually none of the EFTA governments were willing to live with the EEA as it was negotiated. By the conclusion of negotiations, all but Iceland and Liechtenstein had put in EU membership applications. Thus, for most EFTA nations, the EEA was to be merely a transitional arrangement. Swiss voters rejected that EEA in December 1992, effectively freezing their EU application. Norwegian voters rejected EU membership but implicitly accepted the EEA. The EEA thus consists of the EU15 along with Norway, Iceland and Liechtenstein.

In 1997 Switzerland initialled a bilateral accord with the EU after six years of intermittent talks and it should enter into force in 2001. This bilateral deal essentially extends EEA rights and obligations to Switzerland.

As far as the liberalization of TBTs is concerned, the lessons from this history are twofold. First it is not easy to liberalise deeply TBTs among nations, even among rich industrialized nations that share similar views on the subject of regulation. Moreover, when a new nation gains access to an existing arrangement the likely result is a lopsided deal in which the newcomer is forced to accept an immutable status quo. Second, deep TBT liberalization requires institutions that permit surveillance, enforcement and adjudication.

TBT Liberalization in the GATT/WTO

Compared to the EU's active and ongoing efforts to reduce TBTs and avoid new ones, TBT liberalization in the WTO has been passive, aimed primarily at avoiding the most obvious protectionist misuse of standards and regulations. These disciplines, which have evolved over the decades are contained mainly in the so-called TBT Agreement that was part of the Uruguay Round.[20] These apply to standards and regulations governing both products and production processes. There are five main disciplines, including a most favored nation (MFN) obligation for import restrictions, national treatment once the good has entered, the sham principle, the least-restrictive-means principle, and the transparency principle.

MFN and national treatment require no comment here. The sham principle states that standards and regulation should not be "a disguised restriction" on international trade. The least-restrictive-means principle states that regulations should accomplish their regulatory goal by means that are the least restrictive to trade. As part of this, there is a general preference for performance standards as opposed to standards based on product characteristics. There is also a "requirement" to use international standards, but this is largely nullified since the requirement does not apply when international standards would be an "ineffective or inappropriate" means of fulfilling legitimate regulatory objectives. The transparency principle takes several forms. There is an obligation to go through a notice-and-comment period when introducing new measures that may affect trade. Signatories should also maintain an information clearinghouse for standards-related inquiries. Additionally there are a number of hortatory elements such as the one urging signatories to give "positive consideration" to unilaterally recognizing foreign standards and regulations.

It is difficult to judge the liberalizing impact of the TBT Agreement, but few would claim that it has had a major impact. Compliance with the TBT

Agreement has been spotty, outside of OECD nations (Stephenson 1997 p.42). TBTs have, however, been the subject of many disputes. Of the 48 requests for consultation to the WTO's Dispute Settlement Body, 11 concerned standards or invoked the TBT agreement. Another measure of the TBT Agreement's lack of success is that much of the energy and momentum in TBT liberalization has developed outside the WTO framework, in regional arrangements (EU, NAFTA, APEC, US-EU MRA, and so on) and on a sectoral basis. According to Wilson (2000), there is "little evidence to indicate that the WTO Committee on TBTs has influenced the debate."

The International Organization for Standardization

There are three main international standard-setting organizations, the International Organization for Standardization (ISO), the International Electrotechnical Commission (IEC) and the International Telecommunications Union (ITU). The ISO and IEC are private, non-governmental agencies in which each member country designates its own representative. In the most important of the three, the ISO, most members are the private or public national standards bodies from member countries. Development of standards is a lengthy process involving voluntary participation and a consensus decision making rule (Stephenson 1997). Member nations are not obliged to implement the adopted standards. Moreover, compliance to an ISO standard does not bring with it automatic market access in other countries.

According to Stephenson, European members have dominated the international standardization process. For example, European members hold two-thirds of the ISO secretariats (roughly equivalent to a chair). The United States, by contrast, participates in almost all of the ISO's work but holds only 13 percent of the secretariats. Stephenson (1997 p.51) claims that the predominance of Europeans in this process derives from the fact that there are 18 European countries with long-established and sophisticated national standards bodies that are well acquainted with international standardization. Most of these co-ordinate their actions under the aegis of the EU.

While the ISO and its related organizations have an important role to play in TBT liberalization, their actions are essentially limited to areas where sector-by-sector negotiations can lead to mutually acceptable improvements. The ISO has produced about 10,000 standards in its 50-year history. While this is a big number, it is small compared to the 100,000 or so standards that exist today in the United States and EU (Stephenson 1997 p. 47).

The EU-US MRA

In June 1997, the European Union and the United States signed a mutual recognition agreement that deals with the trade-inhibiting aspects of conformity assessment. The MRA commits each party to recognize the results of product testing or certifications of both governments in the six specified sectors. This should liberalize trade in these sectors by reducing the cost of duplicative testing, inspection and certification. The covered sectors are telecommunication equipment, electromagnetic compatibility, electrical safety, recreational craft, pharmaceutical good manufacturing practice, and medical devices.

Importantly, the agreement is confined to conformity assessment and does not mutually recognize standards and regulations themselves. Article 4 of the MRA states, "This Agreement shall not be construed to entail mutual acceptance of standards or technical regulations." It merely permits some EU standards to be assessed by U.S.-based laboratories and some U.S. standards to be assessed by EU-based laboratories. EU and American industries, nevertheless, viewed this, as an important step to reducing trade barriers. Indeed, the initiative came mainly from industry-to-industry talks in the Trans-Atlantic Business Dialogue.

A number of other MRAs have also been signed. One of the first outside Europe is the Joint Accreditation System – Australia and New Zealand (JAS-ANZ). This is an open system that promotes mutual recognition of testing in its member nations and beyond. As part of its mandate, JAS-ANZ facilitated the negotiation of an MRA with the EU.

Future Liberalization

I have argued that TBTs are important and studied the main liberalization efforts. The next step is to argue that TBT liberalization will continue and it will almost surely proceed along the lines of mutual recognition agreements among rich nations, augmented by some unilateral harmonization by poor nations to the standards of rich nations.

The argument requires a bit of background. First, it is necessary to clarify our thinking on how TBTs are different from conventional trade barriers, such as tariffs and quotas. Second, a political-economy framework is needed to help us understand historical liberalization efforts and to predict the path of future TBT liberalization.

Obscurity not Sovereignty

It is often asserted that liberalization of behind-the-border measures – product norms being a prime example – intrinsically engages issues of sovereignty that did not arise in tariff-cutting talks (Vogel, 1997, Bagwell and Staiger 1998, Brittan 1997). This is misleading.

Tariffs, like many TBTs, are adopted in pursuit of governance goals, such as income distribution, that the community of nations views as legitimate. However, via the GATT/WTO, the same community rejects tariffs (or more specifically the raising of bound rates) as a legitimate means of attaining such goals. The essential reason is that tariffs are not the least-restrictive means of fulfilling the goals. Developed nations can achieve the same goal with income transfers, active labor market policies, and the like.[21] The point here is that tariffs are not prohibited because nations reject income distribution as a legitimate policy goal. Tariffs are prohibited because equivalent but less-trade-restrictive measures clearly exist.

This brings us to the key distinction between tariffs and TBTs as far as liberalization efforts are concerned. TBTs are harder to liberalize because it is considerably more difficult to determine whether less-trade-restrictive measures could accomplish the same governance goals. In short, the difference between regulatory protection and tariffs boils down to issues of obscurity, not sovereignty. This simple but important point has significant ramifications for the nature of future liberalization.

Negotiated Harmonization Does Not Work

TBT liberalization can proceed in two ways, mutual recognition and harmonization, with harmonization coming in two basic 'flavors', negotiated harmonization and hegemonic harmonization. Here I argue that negotiated harmonization does not work.

Begin with the historical evidence. The European Union in the 1970s was a nearly ideal setting for such an approach. EU members had a 'constitutional' obligation to approximate product norms, made regular side-payments via the EU budget, had a super-national judicial system for dispute settlement, and had similar income levels and regulatory goals. Even in this setting, negotiated harmonization by unanimity proved unworkable in all but a handful of cases. This suggests that prospects are nil for negotiated harmonization in the wider world for most products.

More can be learned by studying *why* the EU failed at harmonization in the 1970s. The failure took the form of interminable delays (in one famous example, it took 11 years to set norms for mineral water). Why the delay? International negotiations must strive for a balance of commercial gains since each government must align a political consensus behind the final liberalization package. Moreover, when it comes to product norms, all governments must be convinced that the synchronized norm meets their governance goals. Obscurity renders both tasks enormously difficult. It is difficult and time-consuming to determine the commercial impact of each proposed norm. Further complexity is added by the need to determine whether each proposed norm permits an appropriate level of regulatory protection in each nation. Given that a typical international negotiation involves many proposals and many modifications of national norms, obscurity can quickly render such negotiations impractical. In fact, obscurity-induced delays are important enough to make the negotiated harmonization approach a non-starter.

Mutual Recognition

This problem leaves hegemonic harmonization and mutual recognition as the only routes forward and both routes are currently being pursued. Hegemonic harmonization is the default option for many small nations that are heavily dependent on a large trading partner, such as the EU. The European agreements and the EU-Swiss bilateral accords, for instance, commit nations to observing the EU's standards and regulations *en masse* and without substantial modification.

Mutual recognition, or more precisely mutual recognition teamed with some 'new approach' harmonization in sensitive areas, is the standard procedure inside the EU. It is also the liberalization route adopted in the bilateral trans-Atlantic trade talks, APEC and other regional arrangements.[22] Interestingly, the most important MRA pursued to date – the US-EU MRA – is limited to mutual recognition of conformity assessments, not of norms themselves.

Political Economy of TBT Liberalization

Good governance demands product regulation, so the existence of standards and regulations is no mystery. But why do these norms inhibit trade? Protection does not result from random acts of nature and trade liberalization is not the outcome of heroic deeds by high-minded civil servants and political leaders. Protection is endogenous. Proposed protection 'clears' the political market by equalizing the demand and supply of protection.[23] Regulatory protection, like most protection, arises from asymmetry in organization and information costs, which implies that industries organize more effectively than consumers. Norm-setting agencies are thus 'captured' by domestic firms in the regulated industry. The obscurity factor greatly enhances the ease of capture since most industrial standards and regulations are off the radar screens of voters and politicians. Obscurity also shapes the mode of capture. In many nations and industries, domestic firms are the *de facto* authors of their own norms. This is easy to understand.

When regulating a technical field, such as elevators, obscurity abounds. The government, which probably does not employ many full-time experts on elevator manufacturing, asks the opinions of domestic elevator-producing firms. With an eye to their foreign competitors, these firms quite naturally push for norms and conformity practices that raise the cost of imported goods more than the cost of locally produced goods. For the domestic firms this is a straightforward cost-benefit exercise. Domestic firms incur the costs of influencing the norm-setting process when these norms could create sufficient rents to cover those costs. This is not to call domestic firms greedy philistines. Most firms are convinced of the superiority of their own products. Crafting standards to reflect characteristics available only in local products may strike the firms as a way of promoting the public interest. Given the obscurity factor, it is difficult, maybe even impossible, to separate the protectionist and public-interest content of a particular norm.

A good example is the European paper industry. Sweden and Finland produce paper mainly from new trees while French and German producers use large quantities of recycled paper and rags. In the early 1990s, the EU considered a regulation that would require paper boxes sold in the EU to have a certain fraction of recycled paper in them. While this sounded like a 'public interest' sort of regulation, it would have had the effect of eliminating the resource-based advantage of Swedish and Finnish firms, much to the joy of the French and German paper industries. Indeed, it is not clear which production method is more environmentally friendly. Recycling paper requires energy and extensive inputs of chemicals that may be released into the environment while growing more trees in the uninhabited north is inherently a green activity. This counter-argument, however, was not pushed by French and German paper producers. Finland and Sweden ultimately joined the EU, so the regulation was not adopted, but this example illustrates the subtle mixing of public interest and protectionism that the obscurity of technical standards permits. It also illustrates how regional norm setting can subtly create barriers for outside nations.

Given that the level of protection is set in a political market, trade liberalization poses a puzzle. Why should a nation find it politically advantageous to remove barriers that it previously found politically optimal to impose? This is especially puzzling since liberalization has occurred on very regular basis over the last 50 years via multilateral trade negotiations.

The Juggernaut Model of Reciprocal Trade Liberalization

Baldwin (1994 Chapter 3) proposes a solution that focuses on exporters as the key anti-protection force and import-competitors as the key pro-protection force. Announcement of reciprocal trade negotiations, in which better foreign market access is bought with domestic liberalization, starts the juggernaut rolling by multiplying the ranks of pro-trade forces. Exporters, in search of foreign market access, shrug their normal indifference to domestic protection and don the cloak of anti-protectionists. The same goes on in other nations, so the political market for protection clears at a lower level implying some liberalization in all participating nations. With the phasing in of the liberalization, export interests get stronger as they expand output and employment. Import-competitors get weaker as they scale back or shutdown. A few years down the road, another round is launched and the juggernaut rolls on. Barriers are initially lower, but anti-trade forces are weaker and pro-trade forces stronger than they were at the conclusion of the last round, so barriers again come down. The endpoint of this process is zero tariffs in the markets of nations engaged in periodic reciprocal trade talks.

Application to TBT Liberalization

Consider next the application of this model to TBT liberalization. In the political-economy approach, changing the level of regulatory protection requires a change in the forces that determined the political equilibrium. In particular, liberalization has to offer something to domestic firms that they find more attractive than the *status quo*. A tried-and-true method is to engage nations in an exchange-of-market-access negotiation. The political economy reasoning, however, differs from the model exposed above.

Most of the TBT liberalization seen so far has occurred among nations and within industries where the exporting and import-competing firms are one and the same – that is to say, in trading relationships dominated by intraindustry trade. We must therefore modify the juggernaut model to allow a different set of political actors. When it comes to TBT liberalization, the key distinction will be between larger, more efficient firms and smaller, less efficient firms. Two-way regulatory protection results in fragmented markets where firms are dominant in their home market while being marginal players in other markets. This market fragmentation reduces competition, raises profit margins and thus keeps too many firms in business and keeps prices too high. Tearing down these barriers reduces the fragmentation of these markets, producing a pro-competitive effect that puts pressure on profits, especially those of the inefficient (and therefore small) firms. The end result is an industrial restructuring that involves weaker firms merging, exiting, or getting bought up. In the end, a more efficient industrial structure emerges with fewer, bigger, more efficient firms competing more effectively with each other.

The key result here is that two-way liberalization of TBTs systematically favors profits of larger firms. Intuitively, big firms gain both from the liberalization and the exit of the small firms. Given this, the juggernaut logic applies directly with large firms in the role of the main pro-liberalizers and small firms in the role of the main anti-liberalizers. Of course, a negotiating framework based on reciprocity is again of the essence. Large and small firms would be united in their opposition to unilateral liberalization.

With all this on the table, it is a simple matter to predict that large multinational firms will continue to press for two-way liberalization of TBTs and this liberalization is most likely to take the form of preferential arrangements among rich nations.

A Two-tier World Trade System?

Most liberalization of regulatory protection in the past has been bilateral or plurilateral and based on the MRA approach. Future liberalization is likely to continue being so for the reasons discussed above. Yet the MRA approach requires a high level of trust in a partner nation's governance capacity. Since few developing nations are able to generate such trust, the ongoing liberalization process will almost surely exclude them. A two-tier system of market access may thus emerge, with developing countries occupying the second tier. Note that this proposition is not entirely based on abstract reasoning. The EU's Single Market program liberalized TBTs on a preferential basis, which implied that non-EU Western European nations had second-class access to EU markets. Outsiders, such as the EFTA nations, had to scramble to redress the situation by joining the EU or the EEA.

Two features make this two-tier market access pernicious: escalation and the magnification effect. The magnification effect mentioned earlier emphasises the tendency for a general lowering of natural and man-made trade barriers to make the remaining barriers and discrimination more important, not less important.

Escalation concerns the link between product sophistication and the cost of TBTs. Product regulation applies to all goods ranging from apples to jet engines. Generally speaking, however, these norms become more restrictive, complex and costly as one moves up the product-sophistication scale. What this means is that preferential TBT liberalization has an 'escalating' effect on the level of discrimination facing third nations. The analogy with tariff escalation should be clear. Even now, OECD nations tend to have higher tariffs on industrial goods than on raw materials, apart from food. In the 1970s, when OECD tariffs were more important, this pattern of protection had the effect of distorting developing nations' post-tariff comparative advantages away from industry.

There are several things wrong with a two-tier system. First, it is undisciplined. The TBT Agreement in the WTO contains no explicit strictures on the form of MRAs or on their discriminatory effects on third nations. Second, it violates the non-discriminatory and multilateral spirit of the WTO and this, in turn, tends to undermine the rules-based world trade system.[24] For both bilateral MRAs and hegemonic harmonization, the basic principle is 'might makes right'.

Third, it harms the competitiveness of firms located in excluded nations, which would likely be dominated by developing nations. For instance, bilateral foreign direct investment flows clearly show that the EU's Single Market program led to investment diversion (Baldwin, Forslid and Haaland 1997) and trade diversion (Sapir 1997). By the same logic, the EU-U.S. MRAs may well divert investment and trade from nations such as Mexico.

Currently, MRAs outside of Europe concern only testing, and may have a limited impact on competitiveness. If the MRA process gets extended to norms as well as testing, the actual content of standards will be affected. Public choice reasoning would indicate that such an outcome would lead to product norms that disfavor goods made in excluded nations. A good European example of this was the attempt by French and German paper firms to impose a standard that would negate the resource-based comparative advantage of Swedish and Finnish firms. One could similarly imagine the United States and the European Union agreeing to a standard that would offset the comparative advantage of developing-country manufacturers.

Some Proposals

Professors are generally better at identifying and framing policy problems than they are at coming up with workable solutions. Yet even admitting this absolute disadvantage, there do seem to be a couple of basic proposals that would address some of the problems.

Open Liberalization or Article 24 Disciplines

TBT liberalization can have radically different results when it is "open" and when it is "exclusionary" (see Section 3 for the arguments). When it is open, as is the case in the EU's Single Market, TBT liberalization may actually improve the market access of third nations.[25] For example, a Turkish-made skateboard can, in principle, be sold in all 15 EU markets after having been certified in any one. To illustrate the importance of open liberalization, consider what the impact would be if EU's mutual recognition principle were limited to EU-made goods. Recalling that EU nations are allowed to have different standards, third nation discrimination in this counterfactual world could be enormous. EU made goods would have to comply with one norm, while third-country goods would have to comply with 15 different norms.

Generally, exclusion in TBT liberalization is enforced via rules of origin. This observation leads to my first proposal. Regional TBT liberalization initiatives, including MRAs, should not in principle include rules of origin. If they do they would likely be discriminatory and should thus be subject to disciplines like those in Article 24 of the GATT. The point is simple. Rules of origin in a free trade agreement are essential to prevent tariff-evasion. However, rules of origin make no sense in an MRA. A good, regardless of where it was made, either meets the requisite norms, or it does not. Country of origin should not be an issue. Given this, rules of origin in a TBT liberalization arrangement create unnecessary discrimination. The GATT/WTO allows for such deviations from the MFN principle, but all such deviations are subject to disciplines designed to ensure that the primary intent of the preferential liberalization is liberalization rather than preferential treatment.

I would propose that these disciplines include something like the "substantially all trade" dictate in Article 24. This would clearly be difficult since all existing MRA are limited to a relatively narrow range of industrial goods. Recall that the substantially-all-trade rule was adopted to reduce the likelihood that the *main* motive behind a free trade agreement was to discriminate against third parties. A similar discipline would be welcome for MRAs. For example, one might imagine that the EU and Japan could craft an MRA on the norms and testing of semiconductors that would substantially raise the costs faced by U.S. chip makers in the EU and Japanese markets.
The regulatory protection thus provided might be very appealing in the domestic politics of both the EU and Japan. Of course, the United States is big enough to counter such moves. But what would happen if the United States and the EU signed an MRA in grapes that had the effect of reducing the competitiveness of Chilean grapes? Surely, this is exactly the sort of thing the WTO was set up to discipline.

Additionally, something akin to the test on a custom union's common external tariff should be applied to MRAs, although obscurity of TBTs would make this difficult to apply. The point is that the signatories of an MRA should be forced to examine the impact of their preferential liberalization on third nations. There should be an obligation to pay compensation to third parties if the MRA results in worsened market access conditions. The details of such a stricture would require a great deal of elaboration, but the 'injury test' in antidumping and countervailing duty cases might provide an example of how one might proceed.

It is worth noting that the issue of rules of origin in MRAs could become important. Reportedly, the EU asked for rules of origin in the negotiation of the EU-U.S. MRA but was rebuffed by the United States (Wilson 2000 p.39). The American stance, however, was based on the impracticality of ascertaining 'the' country of origin for sophisticated industrial goods (things today are made nowhere in particular). The impact of such rules on third nations was not an issue for either the European Union or the United States.

GSP-like Treatment

My second proposal concerns policies that could actively seek to offset any bias against the interests of developing nations arising from regional TBT liberalization. The TBT Agreement already contains hortatory statements about providing technical assistance to developing nations. The WTO, or some other organization such as the World Bank or UNCTAD, should more explicitly promote the TBT equivalent of the Generalised System of Preferences (GSP). For instance, industrial nations might directly or indirectly subsidise conformity assessment procedures for products made in developing nations. Right now, the TBT Agreement has disciplines against members charging foreign firms certification fees that are too high. To offset the cascading effects of discriminatory TBT liberalization, the WTO could encourage subsidization of fees charged to firms based in developing nations.

Summary

In this chapter, I have covered the economics of technical barriers to trade (TBTs), and presented evidence on their importance and various initiatives made to liberalize them. Most liberalization of regulatory protection in the past has been discriminatory and based on the MRA approach. Future liberalization is likely to continue down this path. The MRA approach requires a level of trust in a partner nation's governance capacity that few developing nations are likely to be able to provide. Consequently, a two-tier system of market access is likely to emerge, with developing countries occupying the second tier.

I also pointed out that the discriminatory liberalization of regulatory protection, unlike preferential tariff cutting, is largely undisciplined despite violating the WTO's spirit of non-discrimination. This lack of discipline may undermine the rules-based system as TBT liberalization becomes increasingly important. More directly, MRAs, especially when they employ rules of origin to exclude third nations from the liberalization, can harm the competitiveness of firms located in excluded nations.

I put forward proposals that the WTO be modified to discipline regional TBT liberalization. Specifically, bilateral and plurilateral TBT liberalization schemes should not include rules of origin. Rules of origin exclude third nations from the liberalization without serving a legitimate purpose. If nations choose to include rules of origin, then the liberalization schemes should be presumed to be discriminatory and be subject to Article 24-like disciplines.

Finally, some sort of GSP-like policy to offset the impact on developing nations should also be considered.

NOTES

1. See Baldwin and Martin (1999) for a detailed comparison of current and pre-WWII levels of openness.

2. See Baldwin and Venables (1995) for more on the economics of preferential liberalization.

3. The existing estimates are based on informal methodologies and often undertaken by interested parties. For instance, an OECD (1996) study roughly calculates the cost of TBTs to be 2-10% of firms' production costs. The ITU found that duplication in US and European testing and certification for telecommunications equipment and information technology products cost industry and consumers over $1.1 billion per year.

4. In the figure, I have assigned the non-preferential dummy coefficient to EC-EFTA trade prior to the 1973 EC-EFTA trade accords.

5. In the diagram the acronym EC is used for both the European Community and the European Union.

6. Most EFTA-EFTA trade is among adjacent nations, such as among Nordic nations or among Alpine nations.

7. Liberalization additionally involves issues of surveillance, enforcement, adjudication and the introduction of new standards.

8. See Baldwin (2001) for details.

9. Tariffs on intra-EFTA industrial trade were eradicated by the Stockholm Convention in 1968 and on EU-EFTA industrial trade by the 1974 EU-EFTA free-trade agreements.

10. See Lauwaars (1988 p.152) for details.

11. An exception, which was a precursor to the new approach, was the 'Low Voltage' Directive of 1973. See Lauwaars (1988 p.156) for details.

12. Commission Communication to the Council of 31 January 1985, "Technical harmonisation and standardisation, a new approach." Elements of this emerged earlier in 1973 with the Council's favourable experience in framing the "Low Voltage" Directive, in 1974 with the Dassonville case, and in 1979 with the Cassis de Dijon case. Other elements emerged later, such as the 'global' approach to conformity assessment in 1989. For a succinct summary see www.NewApproach.org.

13. Baldwin (2001) provides more details.

14. See Sykes 1995 for legal analysis.

15. These harmonized standards are prepared in accordance with the "General Guidelines" agreed between the Commission and the mandate issued by the Commission after consultation with the Member States.

16. "A Global Approach to Certification and Testing", EC Commission, 24 July 1989.

17. The idea was first suggested at a meeting of EFTA and EU ministers in Luxembourg in 1984. This produced the Luxembourg Declaration. EEA talks began informally in 1989, continuing more formally in 1990 and 1991. It was signed 2 May 1992 in Oporto, together with an Agreement establishing the EFTA Court of Justice and the EFTA Surveillance Authority.

18. See Baldwin (2001) for further details.

19. EFTA nations participate in Euro-standards bodies, such as CEN and CENELEC.

20. See Chapter 1 of this volume for details.

21. Counterexamples serve to strengthen the point. For developing nations that do not have sophisticated systems for gathering and distributing tax revenue, tariffs are more likely to be tolerated. Moreover, the GATT allows import prohibitions when these are the only practical means of pursing a legitimate goal, such as reducing the consumption of illegal drugs.

22. See Stephenson (1997), Wilson (1999) and OECD (1998) for details.

23. Mancur Olsen is the Freud of this literature. See Hillman (1989) for the basic logic, and Dixit, Grossman and Helpman (1997) for recent refinements.

24. Note that the TBT Agreement actually encourages MRAs, so the current course of liberalisation is by no means GATT-illegal.

25. The Cassis de Dijon mutual recognition principle applies to any good that is made or marketed in an EU nation.

REFERENCES

Bagwell, K. and R. Staiger. 1998. The simple economics of labor standards and the GATT, NBER (National Bureau of Economic Research) Working Paper 6604.

Baldwin, R. 1994. *Towards an integrated Europe*. CEPR (Centre for Economic Policy Research). On line at http://www.unige.ch/~baldwin/.

———. 2001. Regulatory Protectionism, Developing Nations and a Two-Tier World Trade System. *Brookings Trade Forum*, forthcoming.

Baldwin, R. and A. Venables. 1995. Regional Economic Integration. *Handbook of International Economics*, Volume III, North Holland.

Baldwin, R. and P. Martin. 1999. Two Waves of Globalisation: Superficial Similarities, Fundamental differences. NBER (National Bureau of Economic Research) Working Paper 6904.

Baldwin, R., R. Forslid and J. Haaland. 1996. Investment Creation and Diversion in Europe with R. Forslid and J. Haaland. *The World Economy,* 19: No. 6: 635-659.

Baldwin, R. E. 1970. Nontariff distortions of international trade, Washington, D.C.: Brookings Institution.

Brander, J. and P. Krugman 1983. A 'reciprocal dumping' model of international trade. *Journal of International Economics*, 15: 313-323.

Brittan, L. 1997. *Globalisation vs. Sovereignty? The European Response*. Cambridge University Press.

Buiges, P., F. Ilzkovitz, J-F Lebrun. 1990. The impact of the Internal Market by industrial sector: The challenge for Member States, European Economy, Social Europe, special edition.

Emerson, M. et al. 1989. The Economics of 1992 : The E.C. Commission's Assessment of the Economic Effects of Completing the Internal Market, Oxford University Press.

European Commission. 1985. Completing the Internal Market. Brussels: Commission the European Communities.

Head, K. and T. Mayer. 1999. Non-Europe: The magnitude and causes of market fragmentation in the EU. Memo. University of British Columbia.

Helliwell, J. 1998. How much do national borders matter? Washington, D.C.: Brookings Institution.

Helpman, E. and P. Krugman. 1989. Trade policy and market structure. Cambridge, MA USA: (MIT) Massachusetts Institution of Technology Press.

Henson, S. 1998. Costs associated with divergent national product standards and conformity assessment procedures and the impact on international trade. In *Regulatory reform in the global economy: Asian and Latin American Perspectives*. Paris: OECD.

Hillman, A. 1989. The political economy of protection. Chur, Switzerland: Harwood Academic Publishers.

Holmes, P. 1997. Non-tariff barriers. In *The economics of the Single European Act*. G. Mckenzie and A. Venables, eds. London: MacMillan Press.

Laird, S. 1999. Patterns of Protection and Approaches to Liberalisation. Working paper.

Lauwaars, R. 1988. The 'Model Directive' on Technical Harmonisation. In *1992: One European Market?*, R.Bieber, R. Dehousse, J. Pinder, and J. Weiler (eds.), Nomos Verlagsgesellschaft. Baden-Baden.

Majone, G. 1994. Comparing strategies of regulatory rapproachement. In *Regulatory Cooperation for an interdependent world*. Paris: Organization for Economic Cooperation and Development (OECD).

Markusen, J. and A. Venables. 1988. Trade policy with increasing returns and imperfect competition: Contradictory results from competing assumptions. Journal of International Economics 2: 299-316.

Matutes, C. and P. Regibeau. 1996. A selective review of the economics of standardisation, entry deterrence, technological progress and international competition. European Journal of Political Economy 12: 183-209.

Moenius, J. 1999. Information versus product adaptation: The role of standards in trade. manuscript. University of California-San Diego.

OECD. 1994. *Regulatory co-operation for an interdependent world*. Paris: OECD.

______. 1996. Proceedings from the Conference on Consumer Product Safety Standard and Conformity assessment: The effect on international trade.

———. 1998. Regulatory reform in the global economy: Asian and Latin American perspectives.

Pelkmans, J. 1984. Market integration in the European Community. The Hague: Martinus Nijhoff..

———. 1989. "Regulation and the Single Market: An Economic Perspective" in *The Completion of the Single Market*. H. Siebert (ed.). Tubingen: Mohr.

Sapir, A. 1997. Domino effects in Western Europe, 1960-1992. CEPR Discussion Paper.

Stephenson, S. 1997. Standards, conformity assessments and developing countries. www1.worldbank.org/wbiep/trade/papers/. Washington, D.C.: World Bank.

Sykes, A. 1995. Product standards for internationally integrated goods markets. Washington: D.C. Brookings Institution.

Vogel, D. 1997. *Barriers or benefits? Regulation in transaltantic trade*. Washington: D.C. Brookings Institution.

Wei, S-J. 1996. Intra-national versus international trade: How stubborn are nations in global integration? NBER Working Paper 5531.

Wilson, J. 2000. The Multilateral and Regional Trade Policy Response to Technical Barriers. Work in progress. www1.worldbank.org/wbiep/trade/papers/.

CHAPTER 4

Standards and Related Regulations in International Trade: A Modeling Approach

Mattias Ganslandt
James R. Markusen

Introduction

Standards and technical regulations are increasingly important in the trade-policy agenda. Perhaps this is because they have become quantitatively more frequent and burdensome or perhaps they simply have become relatively more important as traditional trade barriers were reduced. In either event, there is a case for closer examination and more formal modeling of standards as they affect trade and national welfare.

Standards and technical regulations are often portrayed as barriers to trade that restrict competition in the local economy by raising costs to foreign suppliers. The idea that standards may act as anti-competitive, protectionist devices seems obvious. Yet it is clear that they may also have benefits, not just to domestic consumers but also to foreign suppliers. If a standard certifies a product as effective in meeting such goals as safety, health, or compatibility with complementary inputs, such as the power supply, such certification can raise consumer demand for the imports, possibly resulting in increased profits to foreign firms in spite of higher costs.

The purpose of this chapter is to suggest formal approaches to modeling standards in ways that potentially may be implemented on actual data. For all the ideas developed here, we present numerical results from simple simulation models, involving minimum dimensionality and constructed data. The advantages of this approach are to demonstrate that our ideas are potentially operational and to show clearly the effects of a standard in models where all extraneous influences are eliminated. With respect to the latter task, we use simple numerical general equilibrium models in which countries have identical relative productivities (technologies) across sectors, identical relative factor endowments, and constant returns to scale so as to eliminate traditional comparative-advantage motives for trade. We model a stylized sector with imperfectly competitive firms, which is a valid representation of many manufacturing sectors. It is less appropriate for agriculture, although even here the exporting firms, as opposed to farmers, are often few in number. Regardless of the generality of this approach, the techniques for modeling the standards themselves can be applied to any applied-general equilibrium model. A further advantage of the imperfect-competition approach is that it quickly identifies conflicts or complementarities between producer and consumer interests, both within and between countries.

In the next three sections we use a two-good, two-factor, two-country, general equilibrium model in which countries have identical relative productivities across sectors and relative factor endowments.[1] In general we assume that one country is considerably larger than the other, either because of endowment size or sector-neutral technical superiority. Respectively, these assumptions capture the notions of big versus small countries and rich versus poor countries.

In Section 2 we show how to model a purely cost-increasing standard. Thus, there is no beneficial effect from the standard. Not surprisingly, a standard imposed by one country tends to benefit producer interests at the expense of consumer interests. An important, albeit unsurprising, result is that the small country cannot win a "standards war" in which both countries impose cost-increasing standards on imports, analogous to a tariff war in traditional trade theory. Both producers and consumers could be worse off in the small country. Thus an international code that restricted the use of "purely harassment" standards, such as multiple and duplicative inspection requirements, would aid the small country and may even win support from import-competing producers in small nations.

In Section 3 we consider a standard imposed only by the big or rich country. In addition to imposing a cost on exporters from the small or poor country, however, the standard also creates an increased willingness to pay by consumers in the importing nation for the small country's good. We model this idea as sliding the consumer's indifference curve along a budget line so that, for example, at constant prices the consumer now chooses 2.5 units of each country's good instead of four units of the large-country's good and 1 unit of the small-country's good. We permit prices to change in response to the standard. Simulation results illustrate the tradeoff between the higher production costs and additional sales from greater demand for imports.

The model in Section 4 is identical to that in Section 3 except that we assume the standard creates a fixed cost for the small country's firms. For example, a new design might be required, but once this is created the marginal cost of production is unchanged. The interesting thing about this case is that, when there are multiple firms, there may be multiple equilibria. With a coordination problem (or free-rider problem) among firms, there may be one equilibrium in which no firm exports, and one in which there are substantial exports. Thus, regulations of this type create a role for the government in the non-standard-imposing country either to oppose the standard or to coordinate compliance procedures and compliance costs.

In Section 5 we analyze cases in which a standard changes the complementarity or substitutability between products. An example of a standard that increases substitutability would be one permitting personal computers built on IBM and Apple platforms to run the same software. One result is that producer and consumer interests may conflict, with consumer's favoring a common standard and producers opposing it since the standard raises competition. Regarding complementarity, if a computer and a telephone are made to work together, the consumer's willingness to pay for both might increase. Both consumers and producers could benefit, suggesting that standards making different products work together might receive stronger political support than ones making products more similar.

Because this is a tricky problem, we use a closed-economy model in which there are two sets of firms. The problem lies less with the idea of changing the shape of isoquants and more with interpretation of welfare changes. In a traditional Dixit-Stiglitz formulation, for example, poor substitutability is a source of welfare gains (love of variety), so reducing substitutability reduces welfare, other things equal. This loss is balanced in our model by a pro-competitive effect arising from increased substitutability. Great care is needed in interpreting the former effect.

In Section 6, we consider the idea that standards could segment markets, preventing arbitrage in oligopolistic markets. In the automobile industry, for example, country-specific emission and safety standards prevent traders and consumers from moving products from one market to another. In many situations this might not be an issue for developing countries. But there are instances in which a country could either be helped or hurt by restrictions on arbitrage. For example, if arbitrage were impossible a poor country could benefit if a pharmaceutical company chose a low price there, knowing the product could not be shipped back to the high-income country. But in other cases, possibly in the presence of high distribution costs, pricing strategies might imply higher prices in the small market.

All of the models we develop are coded for the MPS/GE solver algorithm and programs are available on request. These exercises illustrate that the models used here may be implemented numerically with constructed data. However, great data difficulties remain before implementation with actual data would be possible. If a standard exists and, therefore, affects values of observed data, the researcher should estimate a shadow value for the restriction. This value could be either positive or negative depending on whether an increased willingness to pay would outweigh increased costs from meeting the standard. In principle, this can be done in the same way that researchers have estimated the shadow rents from quotas, domestic content restrictions, and so forth.

Standards as Real Trade Costs

In this section, we treat standards simply as additional trade costs. We implicitly assume that firms of one country have made some investment to meet their own national standard but must incur a second investment to serve the foreign market. The standards of the two countries are simply different, neither is presumed to be better or worse than the other. Any exports to a country must meet the importing country's standard to be permitted to enter. An example that might come close is 110-volt versus 220-volt systems for small electrical appliances.

The model assumes there are two goods, X and Y, two factors, labor and capital (L and K), and two countries, home and foreign (h and f). Throughout the paper, we will think of country h as a larger, richer developed country, and f as a smaller, poorer developing country. But, as noted in the introduction, we will assume no comparative advantage in the X versus the Y sector so as to focus on the essentials. Thus, country h may have a equal absolute advantage in both sectors. The Y sector produces a homogeneous good with constant returns to scale and perfect competition, while the X sector produces differentiated goods with imperfectly competitive firms. In order to sort out opposing or complementary effects on producers versus consumers, we assume a relatively short-run analysis in which the number of X firms is fixed in each country.

Let X_{ji} be the supply by a single firm in country j to the market in country i. Welfare in country i is given by

$$U_i = Y_i^{1/2} UX_i^{1/2} \qquad UX_i = A\left[n_h \; X_{hi}^{\beta} + n_f \; X_{fi}^{\beta}\right]^{1/\beta} \qquad \sigma = \frac{1}{1-\beta}$$

where n_h and n_f are the number of firms (varieties) in countries h and f respectively. The parameter σ is the elasticity of substitution between any two X goods.

In this section we do not assume any nesting in the utility function, so that two goods from country i have the same substitutability as one good from i and one good from j. The production function for Y in country i is given by

$$Y_i = L_{iy}^{\gamma} K_{iy}^{1-\gamma}$$

Production functions for X are identical across firms and countries. Output by a representative firm in country i is given by

$$X_i = L_{ix}^{\delta} K_{ix}^{1-\delta} .$$

Factors *L* and *K* are in inelastic supply, so factor-market clearing requires that factor demands add up to total factor supplies in country i.

$$L_i^* = L_{iy} + n_i L_{ix} \qquad K_i^* = K_{iy} + n_i K_{ix}$$

Firms in the X sector are imperfectly competitive. Let p_{ij} denote the price for X of a firm producing in market i and selling in market j. Firms set markups m_{ij} according to the formula:

$$p_{ii}(1 - m_{ii}) = mc_i(w_i, r_i) \qquad p_{ij}(1 - m_{ij}) = (1 + s_j) mc_i(w_i, r_i)$$

where mc_i is the marginal cost of producing a variety X_i (an identical function across countries), and *w* and *r* are the prices of labor and capital, respectively. The variable s_j is the cost of meeting country j's standard when exporting from country i. The Cournot formula for the markup of an individual firm, as worked out by Lopez-de-Silanes, et al (1993), is given by

$$m_{ij} = \frac{1}{\sigma} + \left[1 - \frac{1}{\sigma}\right] \frac{n_i X_{ij}}{n_i X_{ij} + n_j X_{jj}} \frac{1}{n_i} .$$

In the case of homogeneous goods where the elasticity of substitution is infinite, this reduces to the well-known equation that a firm=s markup is its market share.

This model can be implemented numerically as a complementarity problem, yielding a set of inequalities with associated non-negative variables. There are 31 non-linear inequalities in this model, which are solved numerically for various values of the parameters. These inequalities are as follows:

Cost functions (complementary variables: outputs)	Number
Y sector output in each country	2
X sector output in each country	2
Cost of supplying market i from market j	4
Production of utility for consumers in each country	2
Production of utility for entrepreneurs in each country	2
Market-clearing equations (complementary variables: prices)	
Market for good Y	1
Total output of X in each country	2
Supply of X from country i to country j	4
Labor-market clearing in each country	2
Capital-market clearing in each country	2
Income-balance equations (complementary variable: the price of utility)	
Income balance for consumers in each country	2
Income balance for entrepreneurs in each country	2
Auxiliary equations (complementary variable: markups)	
Markup equation for producer in market i serving market j	4

The constructed data set and calibration of the model are available on request. We assume that country h has 4/5 of world income and country f has 1/5. We can think of this as endowment size differences, total factor productivity differences, or some combination. Figures 4.1 and 4.2, respectively, give the effect of a large-country standard s_h on the profits (PROF) and consumer welfare, defined as real income (WCON) in country h and country f. Consumer welfare refers to total factor income divided by the consumer price index. Profits are markup revenues divided by the same price index and we assume that firm owners have the same tastes as factor owners. In the model's benchmark equilibrium, profits are 10 percent of total income.

The standard is modeled as a proportionate increase in marginal production costs, as shown on the horizontal axes. Consumers in both countries would be made worse off by the standard, almost to the same extent in the two countries. Not surprisingly, profits would increase substantially for firms in the large country and fall for the small-country firms. Thus the standard would serve the political interest of the large-country firms at the expense of everyone else.[2]

FIGURE 4.1. Effect of a large-country standard on the large country

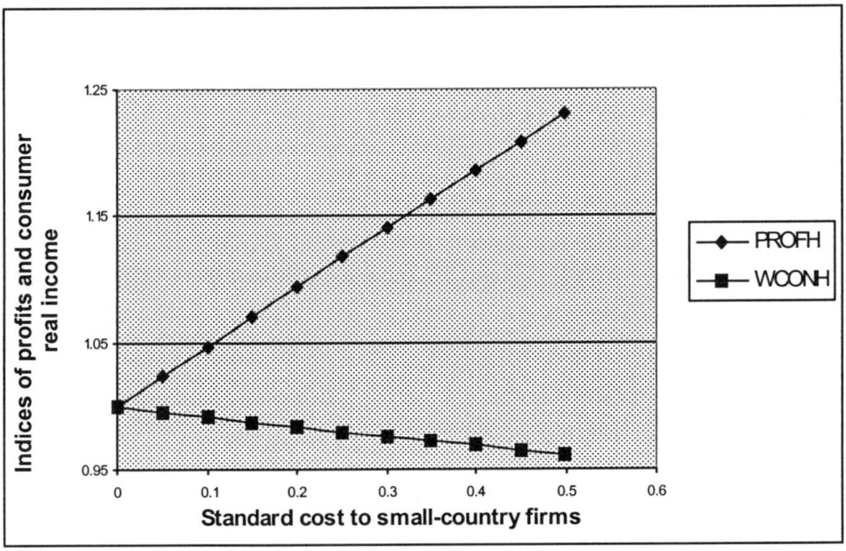

FIGURE 4.2. Effect of a large-country standard on the small country

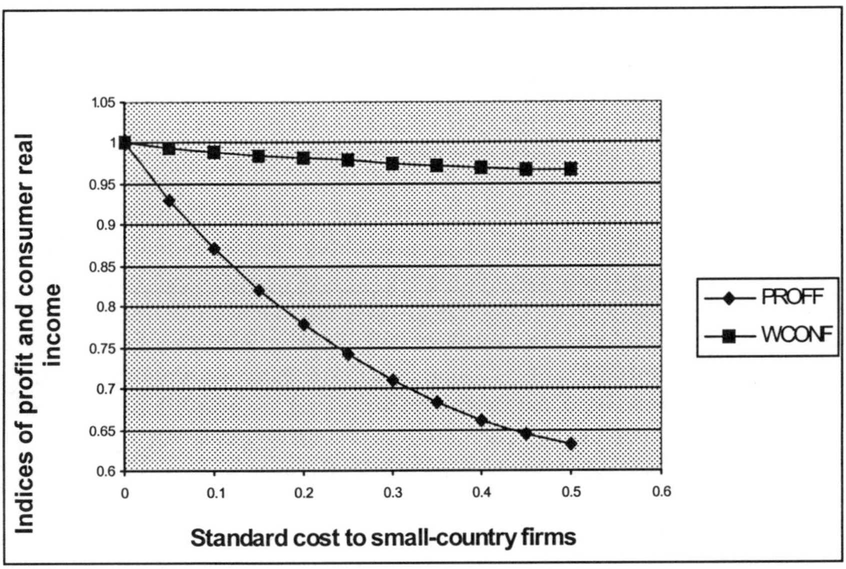

Figures 4.3 and 4.4 show the effects of a standard imposed by the small country f. Results in the top panel are qualitatively similar to Figure 1, but far less significant quantitatively. The small country cannot hurt the large country very much. An interesting result occurs in the bottom panel, where the small-country's standard actually would hurt its own firms up to the level of about 20 percent additional costs. This is a general-equilibrium effect, in that the implied restriction on imports from the large country would cause the latter's firms to supply more to their own market, supplanting imports from the small country.[3] For low levels of the restriction, this effect would reduce profits of the small-country's firms by more than it would raise their profits from protection at home.

Figures 4.5 and 4.6 show what happens when both countries impose the same *ad valorem* restrictive standard. Consumers in both countries would be hurt, but by far more in the small country f (panel B) than in the large country h (panel A). Profits would increase for firms in the large country even though they face an import barrier in country f. Profits for firms in the small country would fall up until a bilateral barrier of 20 percent and then begin to rise, but would recover their free-trade level at about a 50-percent barrier.

The clear message from Figures 4.5 and 4.6 is that the smaller, poorer country cannot win a standards war. Such a war can only benefit firms in the large country. Conversely, an agreement to limit purely cost-increasing standards would help the small country most (although it would help large-country consumers as well). This result accords well with familiar results from the literature on tariff retaliation.

Standards that Affect Consumer Willingness to Pay

In many situations standards are not simply a cost-raising measure without offsetting positive value. Consumers (through importers and wholesalers) place value on accurate certification of such things as health and safety standards. The difficulty is that one could imagine a whole portfolio of models to deal with these issues. Models could involve asymmetric information, markets-for-lemons problems, and so forth. The purpose of this paper is to suggest simple and tractable ways to implement standards into numerical general-equilibrium models. Thus, here we take a simple approach in which we model the "certification value" as an increased willingness to pay by consumers. We refer to the added amount that American consumers would be willing to pay for Mexican tomatoes if they are certified as healthy and safe according to U.S. standards.

Unfortunately, that is not the end of the issue, since there are many ways to

FIGURE 4.3. Effect of a small-country standard on the large country

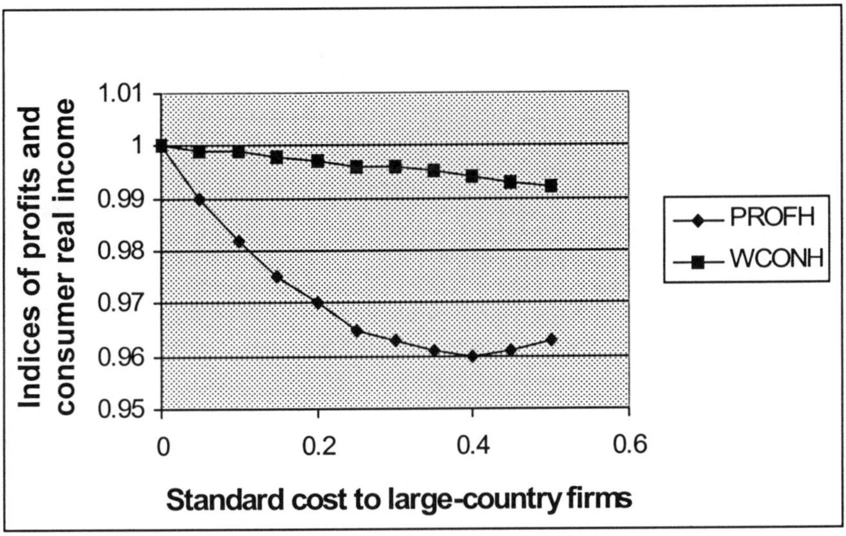

FIGURE 4.4. Effect of a small-country standard on the small country

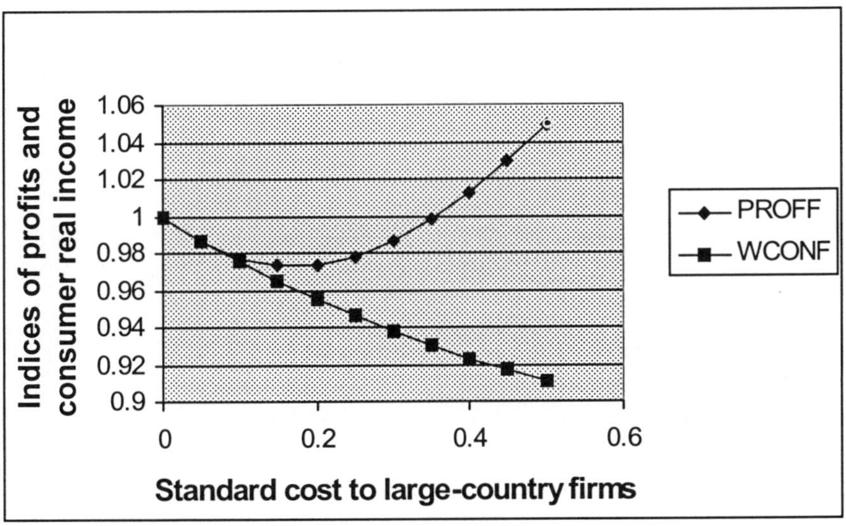

FIGURE 4.5. Effect of bilateral costs on the large country

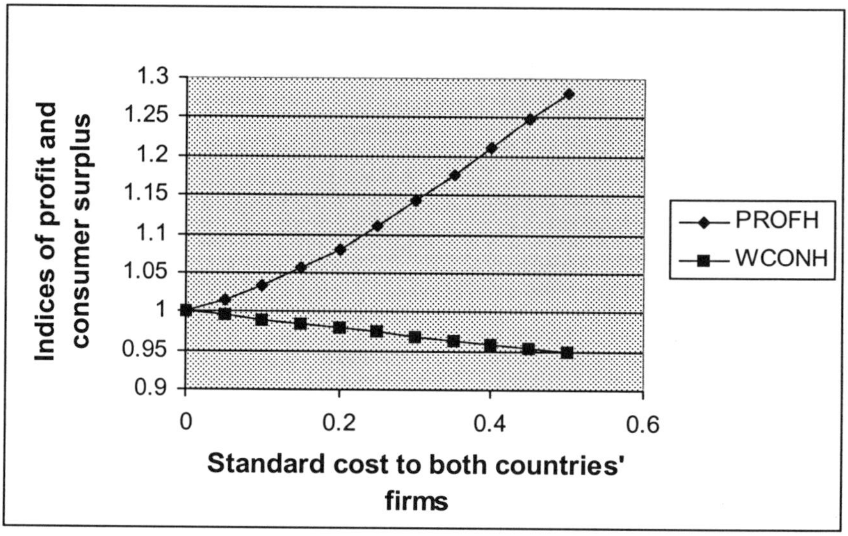

FIGURE 4.6. Effect of bilateral costs on the small country

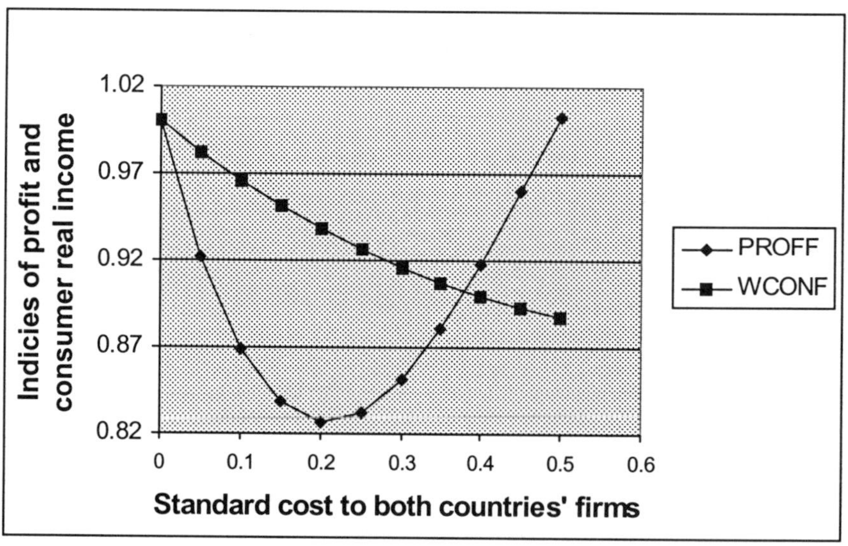

model *formally* an increased willingness to pay. Such a change would alter the utility function, shifting and twisting an indifference curve between the domestic and imported product in some way. In principle, there are infinitely many ways to do this. Considerable care is needed, in that different alternatives have quite different implications for welfare measures.

Figures 4.7 and 4.8 illustrate two possibilities. In Figure 4.7, an increased willingness to pay for X_f is modeled as simply a change in the ratio of X_f to X_h demanded at a constant price ratio, holding utility constant. Both curves UX^0 represent the same utility level. In Figure 4.8, an increased willingness to pay is modeled by the shift in UX^0 (again, the two curves give the same utility level). Consumers now would get the same utility level from less X_f or, alternatively, would be willing to pay more for the same level of X_f. Note that there is a welfare difference between these two conceptions of changed preferences. In Figure 4.8 there is a new indifference curve tangent to the original budget line through X^0 to the northwest of X^0 (if the elasticity of substitution between the two goods is greater than one). Thus, it appears that the second reaction leads to a similar sliding along the original budget line as in Figure 4.7. But if we did this in Figure 4.8, the new indifference curve on the original budget line would correspond to a higher level of utility, not the same level of utility as in Figure 4.7. Figure 4.8. could be thought analogous to a technical improvement, while Figure 4.7 could not. It is not clear which procedure is correct, but the modeler must remember that they are not the same.

In this section, we model an increased willingness to pay as in Figure 4.7, while we use the procedure in Figure 4.8 in the next section. We retain the same model and calibration of the previous section. Assume that only the large country imposes a standard. Figures 4.9 and 4.10 show the effects on profits of the small-country firms and on consumer real income, respectively. The numbers on the "willingness to pay" axis are the share of X purchases by large-country h's consumers that would go to the small-country f's good if the prices of the two were the same. That is, the axis measures country h's demand share $X_f/(X_h + X_f)$ at equal prices for the two goods (note that equilibrium prices do not stay the same after we recalibrate country h's preferences).

The results are not surprising. They make it clear that there is no conflict between the interests of small-country producers and consumers when it comes to the standard imposed by the large country. Effects on consumers would be much smaller than on producers, which also makes sense.

FIGURE 4.7. Standard increases willingness to pay for X_f, version 1

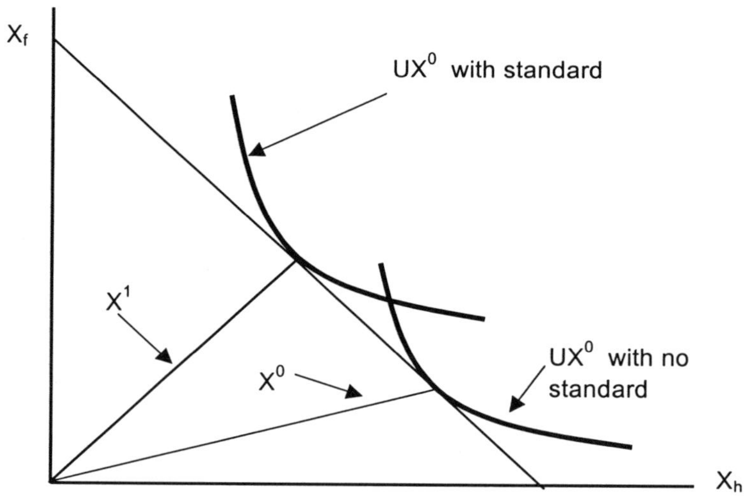

FIGURE 4.8. Standard increases willingness to pay for X_f, version 2

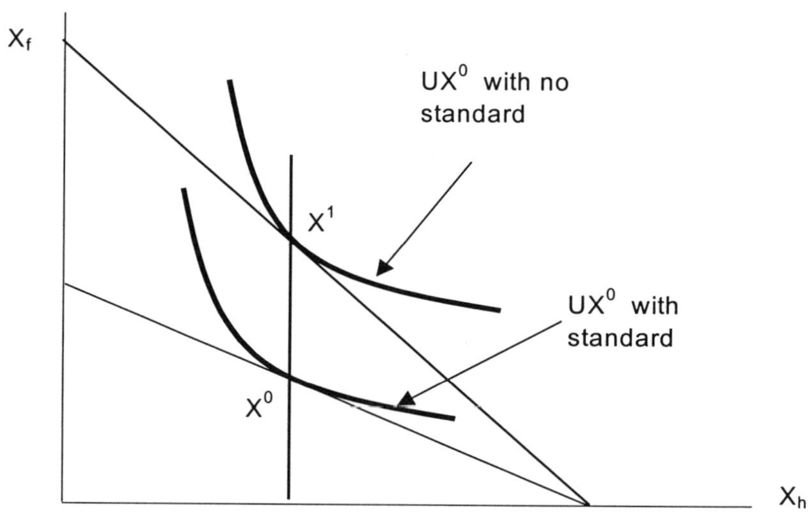

FIGURE 4.9. Effect of a large-country standard on firm profits in the small country

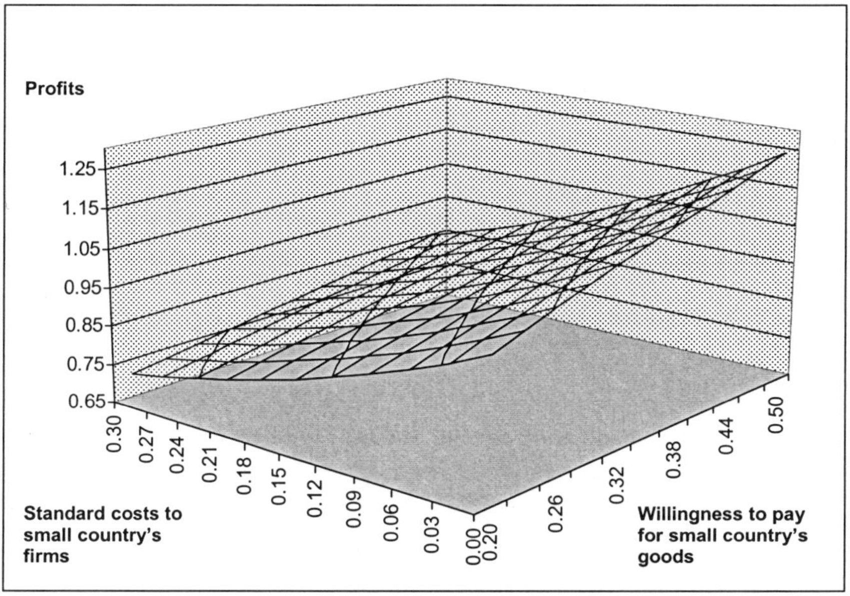

FIGURE 4.10. Effect of a large-country standard on consumer welfare in the small country

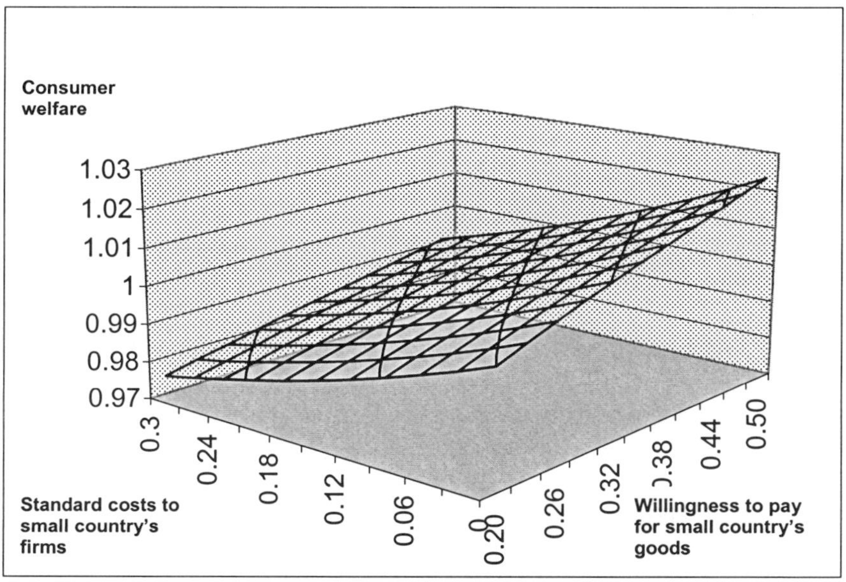

Standards that Impose a Fixed Cost of Entry

In many cases, it is not reasonable to assume that a standard imposes a constant proportional cost. It may require a redesign cost, after which the good can be produced at the old marginal cost. The new design created by this additional fixed cost can be purely private, or it may be a non-excludable public good. The latter case is particularly interesting, and leads to different theoretical, modeling, and policy issues. One of these is the possibility of multiple equilibria, which will be the focus of this section. At a given, fixed (e.g., redesign) costs needed to meet the large-country's standard, there may be two equilibria, one with zero exports from the small country and one with positive exports. There is also likely, a third equilibrium intermediate to these cases as well.

Again, we use the same model and calibration of the previous sections. There is an "endogenous tax rate" on exports from country f and this tax rate is set by a constraint equation. A fixed tax rate is divided by the activity level of exports. This level is equal to one in the initial calibration. As exports go to zero, the ad valorem equivalent of the fixed cost goes to infinity.

Table 4.1 shows results for the model, where the benchmark values in the top row are all normalized to unity. The fixed rate, denoted "cost," is in-

TABLE 4.1. Standard in Country h Imposes a Fixed Cost for Entering the Export Market in Country f (creates multiple equilibria)

	Country h			Country f		
	cons h	profit h	welf h	cons f	profit f	welf f
Benchmark cost = 0 wtp = 1.0	1.000	1.000	1.000	1.000	1.000	1.000
eq. with exports cost = 0.1 wtp = 1.0	0.989	1.060	0.996	0.987	0.843	0.973
eq. with no exports	0.923	1.540	0.985	0.958	0.649	0.927
eq. with exports cost = 0.1 wtp = 1.10	0.997	1.019	0.999	1.001	0.980	0.999
eq. with no exports	0.923	1.540	0.985	0.958	0.649	0.927
eq. with exports cost = 0.1 wtp = 1.25	1.008	0.997	1.005	1.023	1.260	1.041
eq. with no exports	0.923	1.540	0.985	0.958	0.649	0.927

Source: Author's simulation result.

itially zero and the willingness to pay (wtp) equals 1.0. For each country we report consumer real income (cons), profits of the local firms (profit), and aggregate welfare or real income (welf). The willingness to pay is modeled as in Figure 4.8, where the numerical values in Table 1 are the ratio X^1/X^0.

The fixed cost set at 0.1 corresponds to an *ad valorem* rate of 10 percent at the initial activity level of 1.0 for exports to country h. There is no increased willingness to pay. In this case there are two equilibria, the first with positive exports and the second with no exports. The former outcome is roughly similar to the results in Figures 4.1-4.3 except that the export activity level would fall below one, meaning that the *ad valorem* equivalent standard would exceed 10 percent. The no-exports equilibrium is very different. Country h firms would get a large profit boost, while large-country consumers, small-country consumers, and small-country firms would be worse off. Note that aggregate welfare would fall in the large country h as well as in the small country f.

The next two sets of results raise the willingness to pay, first to 1.1 and then to 1.25. There are multiple equilibria in both cases. The no-export equilibrium is the same in all cases, but we have reproduced the numbers anyway for easy comparison with the equilibrium with positive exports. It is interesting to note that a wtp of 1.1 and a fixed cost of 0.1 nearly offset each other in the results for consumers and firms in the positive-trade equilibrium.

The final set of numbers, with wtp set to 1.25, indicate that the standard significantly would improve the welfare of country f's firms and consumers in the positive-exports equilibrium. But the bad outcome, with no exports, remains a possibility. This kind of situation clearly calls for an active policy in the smaller country to aid industry to adapt to foreign standards.

Standards that Alter the Substitutability or Complementarity between Products

In this section we present a simple closed-economy general-equilibrium model in order to examine how standards might affect the relationship between different product types. Suppose that there are two goods, X and Y, and two factors, L and K. The X goods come in two types or standards, X_1 and X_2, these being imperfect substitutes. There may be several firms producing each standard, and the output of standard X_i by firm j is denoted X_{ji}. There is a representative consumer who derives utility from Y and a CES composite of

$$U = Y^{1/2} UX^{1/2} \qquad UX = A\left[X_1^{\beta} + X_2^{\beta}\right]^{1/\beta} \qquad \sigma = \frac{1}{1-\beta} \qquad X_i = \sum_j X_{ji}$$

X_1 and X_2, denoted *UX*. Where σ is the elasticity of substitution between goods of different standards. The production function for Y is given by

$$Y = L_y^{\gamma} K_y^{1-\gamma}$$

Production functions for X are identical across firms within and between types.

$$X_{ij} = L_{ijx}^{\delta} K_{ijx}^{1-\delta}$$

L and *K* are in inelastic supply, so factor-market clearing requires that factor demands add up to total factor supplies.

$$L^* = L_y + \sum_i \sum_j L_{ijx} \qquad K^* = K_y + \sum_i \sum_j K_{ijx}$$

Firms in the X sector are imperfectly competitive. Firms of standard j set a markup m_j according to the formula

$$p_{xj}(1 - m_j) = mc_{xj}(w, r)$$

where p_{xj} is the price of standard j. The term mc_{xj} is the marginal cost of producing x_j (assumed to be identical across standards), and *w* and *r* are the prices of labor (*L*) and capital (*K*). Let σ_w denote the elasticity of substitution within a firm type (e.g., between the products of two type-1 firms) and σ_b the elasticity of substitution between firm types (i.e., between the products of a type-1 and a type-2 firm). The Cournot formula for the markup of an individual type-j firm is given by:

$$m_j = \frac{1}{\sigma_w} + \left[\frac{1}{\sigma_b} - \frac{1}{\sigma_w}\right]\frac{1}{n_j} + \left[1 - \frac{1}{\sigma_b}\right]\frac{X_j}{X_i + X_j}\frac{1}{n_j}$$

where n_j is the number of type-j firms. In our case, goods of given standard are perfect substitutes, so σ_w is infinite. This simplifies the formula to

$$m_j = \left[\frac{1}{\sigma_b} + \left[1 - \frac{1}{\sigma_b}\right] s_j\right]\frac{1}{n_j} = \frac{s_j}{n_j} + \left[1 - \frac{s_j}{n_j}\right]\frac{1}{\sigma_b}$$

where s_j is the share of all type-j firms in total X output. Thus, s_j / n_j is the overall X market share of one type-j firm. The markup is increasing in a firm's market share and decreasing in the elasticity of substitution between standards.

Consider first the situation in which a standard improves the substitutability between products X_1 and X_2. The situation as we will model it is shown in Figures 4.11 and 4.12. Two indifference curves for a given level of UX, denoted UX_0, are shown in Figure 4.11. The curve without the standard has a low elasticity of substitution, while that with a standard has a higher elasticity of substitution. In order for the two curves to correspond to the same value of utility, the constant term A in the good-X sub-utility function must be adjusted to offset the change in β. Thus, if β is increased from β_0 to β_1, the intercept A must be recalculated as follows.

$$UX^0 = A_0\left[(X_1^0)^{\beta_0}+(X_2^0)^{\beta_0}\right]^{1/\beta_0} = A_0\, 2^{1/\beta_0} X^0 = A_1\, 2^{1/\beta_1} X^0 \qquad X^0 = X_1^0 = X_2^0$$

In this equation use was made of the illustrated assumption that X_1 and X_2 are the same quantity. This change in values implies that

$$A_1 = A_0\, 2^{(1/\beta_0 - 1/\beta_1)}$$

The compensating change in A to offset the change in β (σ) means that sub-utility UX^0 can still be produced from equal levels of X_1 and X_2 (X^0). Alternatively, we can think of UX^0 as a final good being produced by inputs, with a higher substitution elasticity between them along an isoquant. We might also believe that in a wide range of specific cases, the standard would affect the cost of production or alternatively the consumers' willingness to pay. Thus the standard may also move the UX^0 isoquant in or out as shown in Figure 4.12. Holding UX^0 constant, movement of this isoquant inward can be thought of as reduced costs of production. Alternatively, we can think of this as an increased willingness to pay, in that the consumer is willing to pay the same amount derived from a smaller quantity of inputs.

We can represent either case as a further change in A. Thus, a standard would have an effect on β (σ) and A. The change in A would come in two parts. The first compensates for the change in β (σ) so that output UX^0 could still be produced from equal inputs of X_1 and X_2 (X^0). Then a subsequent change in A would shift the UX^0 isoquant in or out as in Figure 4.12. From this point on, we refer to only this second impact as the change in cost or change in willingness to pay.

An important point in this model is that the increased elasticity of substitution between X_1 and X_2 would lead to a pro-competitive effect via a decrease

FIGURE 4.11. Standards improve the substitutability between products

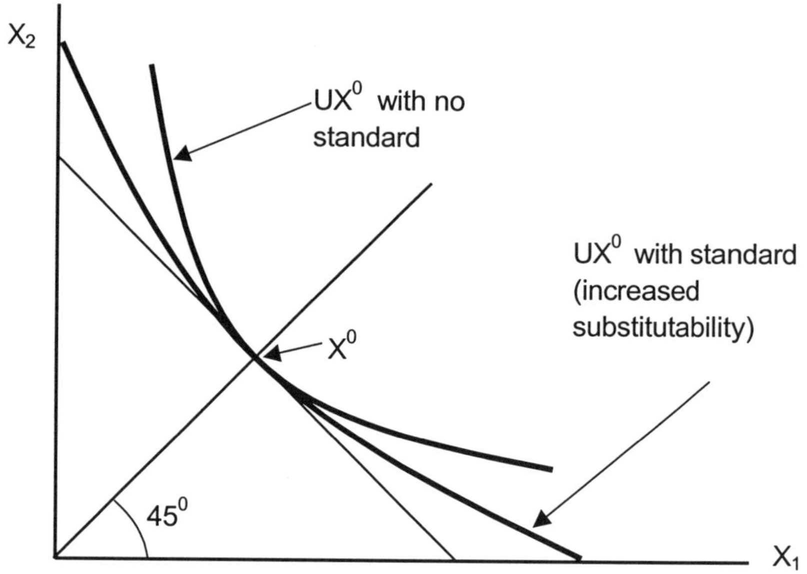

FIGURE 4.12. Standards also alter cost or willingness to pay

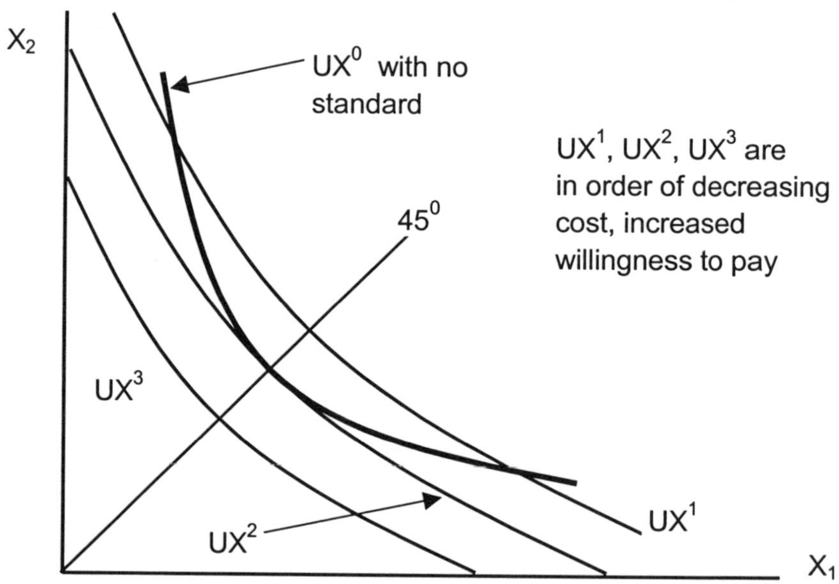

in markups. Referring back to the markup equation, we see that an increase in β (σ) would promote to a fall in the markup (*s/n* must be less than one). By assumption the two product groups are symmetric, meaning that *s* is constant at 1/2 when, as here, we hold the number of firms (*n*) fixed. With no change in willingness to pay, this should imply an increase in consumer welfare and a fall in producer profit. Indeed, simple intuition suggests that overall welfare should rise with the increase in substitution between products.

In other situations we might expect that standards should make products better complements. Two products might have different functions, but if a common standard allowed them to work together then consumers may be more inclined to buy both. If telephones and computers can be made to work together, they would become complements. This situation is modeled in Figures 4.13 and 4.14. The indifference curve (or isoquant) UX^0 with no standard has a relatively high elasticity of substitution. If the two products can be made to work together (e.g., the compatibility of the phone system and computers allowed for the development of the internet), then they become complements. Specifically, assume that there is no change in the consumer's willingness to pay for *either* a phone or a computer separately. Thus, the endpoints of the UX^0 indifference curve in Figure 4.13 would not change. But the consumer may now be willing to pay more for *both* products together. In Figure 4.13, the point on the 45^0 line with the standard (X^1) would lie inside the original no-standard point X^0. Let $\alpha < 1$ denote the ratio X^1/X^0 in Figure 4.13 . A given α requires a re-calibrated elasticity of substitution. The intercept of the production function *A* would not change because the with-standard and without-standard curves go through the same points on the axes). For a given α, the new value of β (σ) is found by solving the following equation.

$$UX^0 = A\ 2^{1/\beta_0} X^0 = A\ 2^{1/\beta_1} \alpha\ X^0 \quad \text{subject to} \quad X^0 = X_1^0 = X_2^0$$

The solution is

$$\beta_1 = \left[\frac{1}{\beta_0} - \frac{\ln \alpha}{\ln 2} \right]^{-1} \qquad \sigma_1 = \frac{1}{1 - \beta_1}$$

Thus a higher α (X^1/X^0) would imply a lower β (σ) in Figure 4.13.

As in our previous example, the standard may also change the consumer's willingness to pay or the costs of production. Thus we might again want to consider changes in *A*, shifting the quantities of X_1 and X_2 necessary to achieve level UX^0. This is shown in Figure 4.14. The UX^2 isoquant corresponds to no change in *A*.

Referring back to the markup equation, we see that increased complementarity should raise the equilibrium markup. Thus the changes shown in Figures 4.13 and 4.14, if there were no change in willingness to pay, should improve producer profit. We would also expect an increase in consumer welfare, although a strong markup effect could reverse this. Barring the latter outcome, consumer and producer interests could coincide in favor of the standard, unlike the increased substitutability case with no change in willingness to pay.

Figures 4.15 and 4.16 present simulation results for this closed economy general-equilibrium model of increased substitutability. On one axis we plot the elasticity of substitution (σ) between X_1 and X_2. The elasticity in the absence of a standard is assumed to equal 0.5, reflecting the idea that the goods are poor substitutes without standardization. The other axis plots willingness to pay, the parameter *A* above, where the no-standard value is 1.0. Alternatively, we can think of this as an inverse cost measure, where the standard would raise production costs. We allow for the possibility that the standard would increase cost (decrease willingness to pay), as was illustrated by isoquant UX_1 in Figure 4.12. The model is calibrated so that there are five firms producing each of the X inputs.

Consumer welfare (real factor income) is shown in Figure 4.15. over a grid of values for σ and *A*. We discover a sharp increase in consumer welfare as σ rises from 0.5, but this effect tapers off at about $\sigma = 3$. Consumer welfare, on the other hand, would rise in a near-linear fashion as the willingness to pay increases. Figure 4.16 demonstrates firm profits, which would decrease as the elasticity of substitution increases. This effect also largely would disappear above a value of $\sigma = 3$. There would be little effect of a change in willingness to pay on firm profits.

Although we do not show the result diagramatically, we also computed total welfare, or the sum of consumer welfare and profits. We find that total welfare would be insensitive to the elasticity of substitution. Thus, increasing the substitutability between products largely would redistribute income from profits to consumers. This result relies on the assumption of a fixed number of firms, as we discus below.

Figures 4.17 and 4.18 show the case where standards improve complementarity between inputs. One axis maps the ratio X^1/X^0 as discussed in Figure 4.13. Lower values of this ratio correspond a more "bowed-in" isoquant and thus to lower values of σ. The other axis is again willingness to pay (*A*) or inverse cost. It does not seem sensible for this ratio to be quite small, since such a magnitude would imply an increased willingness to pay even for one good alone, and it is not clear why this should be the case. Thus we run this parameter over a more limited range in this simulation.

FIGURE 4.13. Standards improve the complementarity between products

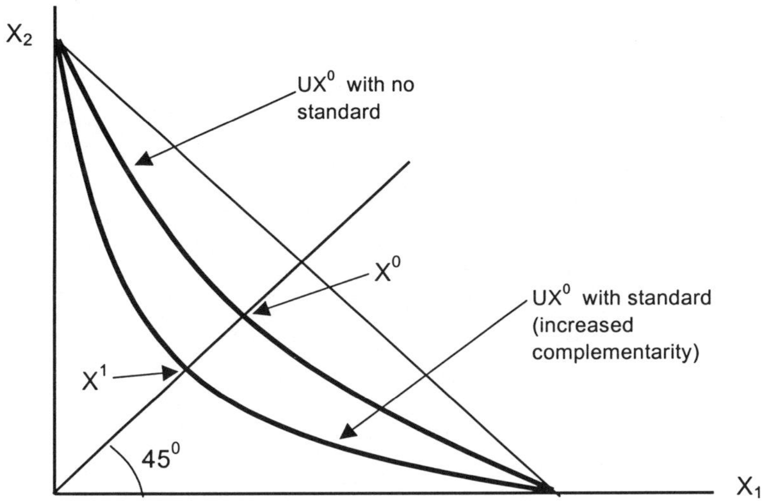

FIGURE 4.14. Standards also alter cost or willingness to pay

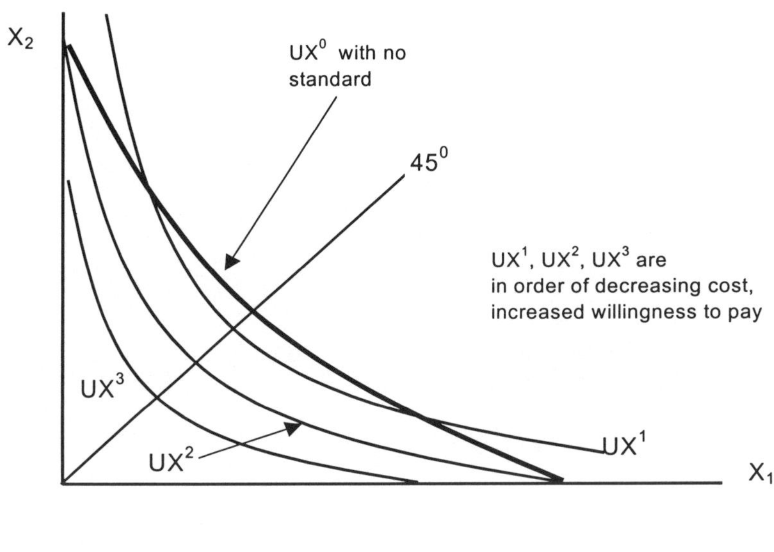

FIGURE 4.15. Consumer welfare, standards improve substitutability

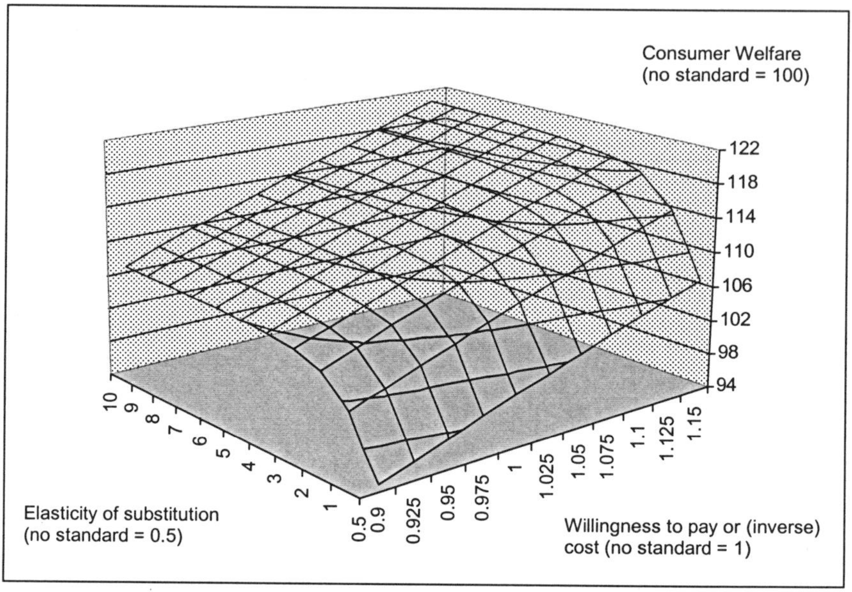

FIGURE 4.16. Firm profits, standards improve substitutability

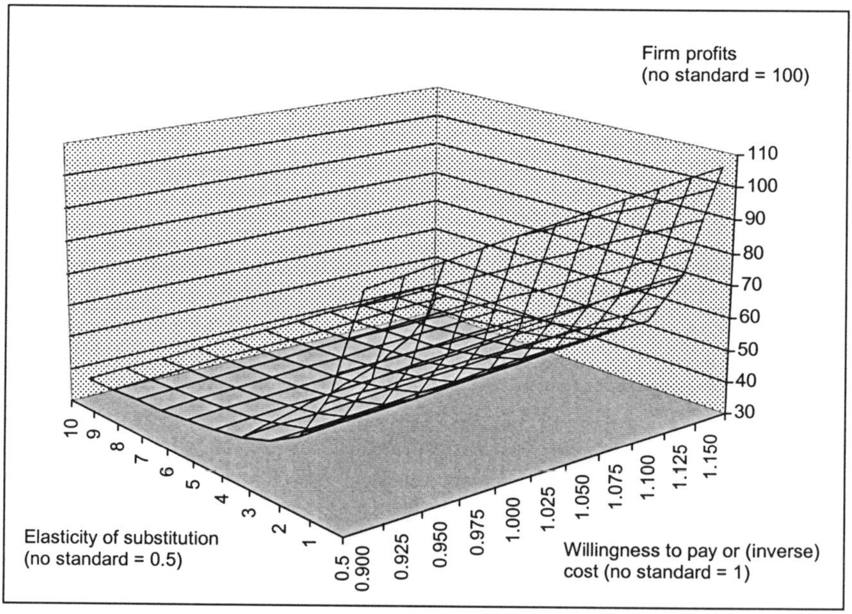

FIGURE 4.17. Consumer welfare, standards improve complementarity

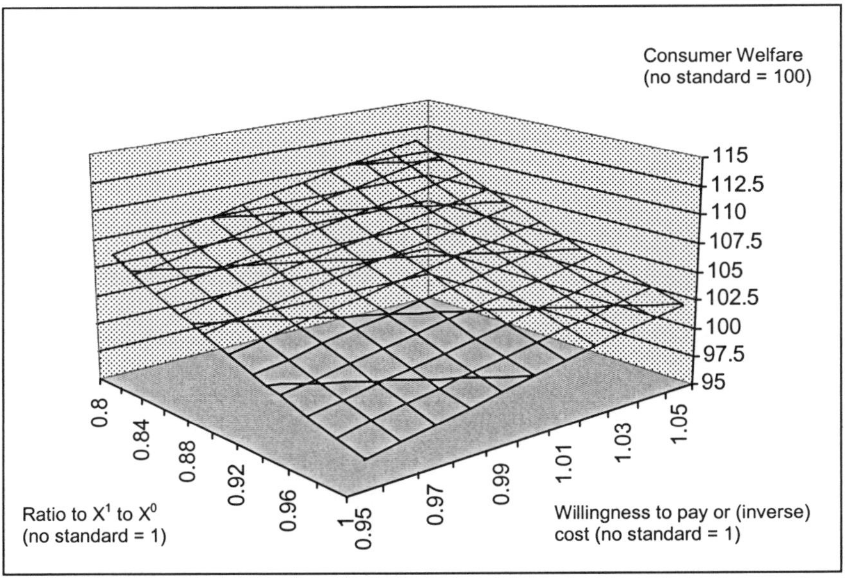

FIGURE 4.18. Firm profits, standards improve complementarity

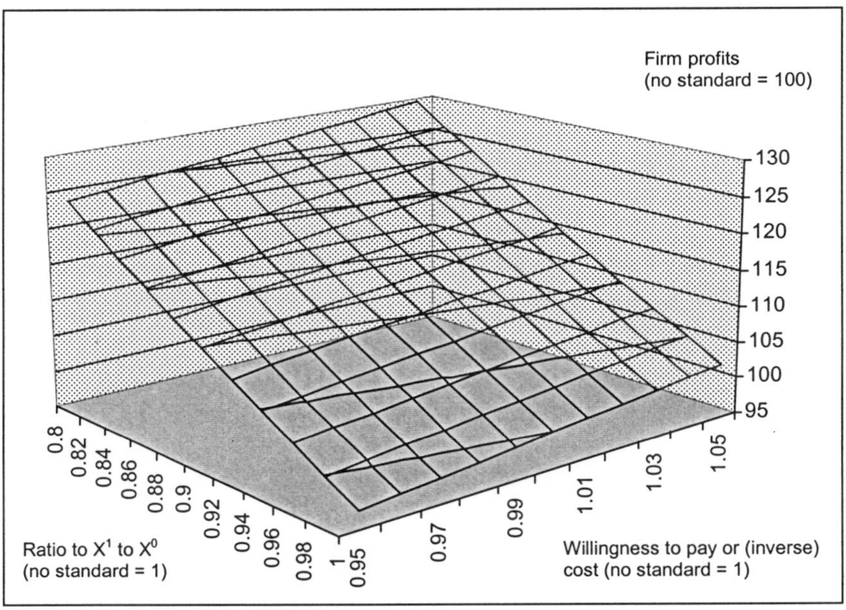

Results in Figure 4.17 show that consumer welfare would rise with a fall in X^1/X^0 and an increase in the willingness to pay. The interesting finding is in Figure 4.18, where we see that profits also would rise with lower values of X^1/X^0. This is due to the fact that these lower values correspond to lower elasticities of substitution between product types and thus support higher equilibrium markups. Total welfare would go up with reductions in X^1/X^0 and increases in willingness to pay. Thus, when standards make goods better complements, we conclude that consumer and producer interests may coincide.

In models with imperfect competition, results often depend on whether there is free entry and exit of firms. Recognizing this, we next add scale economies in the form of fixed costs for each firm in the X sector, and allow free entry and exit to occur until profits are zero. Rather than go into detail here, we merely report the differences this assumption makes. In the case of increased substitutability, free entry and exit would increase the change in welfare due to a rise in σ from 0.5 to 10.0 by a factor of four. The lower markups caused by to the increased elasticity would induce about 40% of firms to exit, leading to an efficiency gain of higher output per firm that would outweigh a mitigating effect of exit on markups. We note that the absolute effect would be small, because the welfare increase would amount to four percent, instead of one percent, of initial income.

In the case of increased complementarity, free entry and exit would reduce the change in welfare coming from a decrease in X^1/X^0. Employing a 20 percent decrease in this ratio in our simulations would result in a welfare gain that is about four percent smaller in the case of free entry and exit as it was in case of a fixed number of firms. This is due to entry of additional firms induced by the lower elasticity of substitution. While this difference between free entry and no entry looks small compared to the case of increased substitutability, the absolute changes would be larger in the present case. In summary, the assumption of free entry and exit would magnify the favorable effects of increased substitutability and reduce the effects in the case of increased complementarity.

Standards and International Arbitrage

Standards have consequences for the costs of trading products and services. In particular, standards may affect the scope for international arbitrage, which has implications for international consumption distortions, entry of new products and the efficiency of production in imperfectly competitive markets. Different national standards can affect either the variable cost or fixed cost of engaging in international trade. These costs can apply to all trade or only arbitrage activities. The exact cost structure depends on border controls, test procedures and the need to adapt foreign varieties to the local market. Moreover, the effect on individual arbitrage by consumers and commercial arbitrage by parallel traders is potentially different. For sufficiently high costs, however, arbitrage is completely blocked irrespective of the type of arbitrage we are considering.

Throughout this section we focus on the arbitrage costs raised by different standards. In contrast to the previous sections, this section analyzes models in which the costs of different standards do not apply to the manufacturing firms' production and distribution of the goods.

Before we study the effects of different standards it should be noted that arbitrage in international markets can be modeled in at least three ways, all of which are applied in the following analysis. One alternative is to derive an equilibrium condition for prices and impose this constraint on the solution. Under this assumption, arbitrage determines the equilibrium but does not occur in equilibrium. To the extent that arbitrage is a costly activity, this approach does not fully take into account the costs associated with price convergence. Another alternative is, therefore, to model arbitrage as an explicit activity. Arbitrage firms (also called parallel importers) buy the goods in markets with low prices and ship them to markets with high prices. Under the assumption of free entry in the arbitrage sector, international price differentials are bounded by the average cost of arbitrage. A final alternative is to focus on two extreme cases, involving either fully integrated or perfectly segmented markets. The former case corresponds to a common standard in all markets and the latter case corresponds to different national standards. This is a convenient way to model different national standards as a prohibitive cost for arbitrage.

In some cases commercial arbitrage is explicitly prohibited. For instance, parallel importers are generally not allowed to exploit international price differentials for patented, copyrighted or trademarked products between developing countries and the European Union. The United States restricts parallel imports of products covered by copyrights and patents. In other cases, arbitrage by single individuals may be prohibitively costly due to increasing returns to scale in the transportation technology, which makes it reasonable to focus on parallel trade by organized distributors.

Standards and International Consumption Distortions

First, we focus on the interaction between standards and "individual arbitrage," which refers to consumers buying the goods directly from a foreign market. As noted, standards can raise the fixed as well as the variable cost of arbitrage. There is no reason to rule out either of these costs *a priori*. In many cases different standards may erect a mix of fixed and variable arbitrage costs. Testing procedures and border controls are mainly fixed costs. Increasing returns to scale in the transportation technology implies high unit arbitrage costs for small traders. Adapting the product to the local market could result in both variable and fixed costs.

We may incorporate both types of arbitrage costs of different standards into a derived individual arbitrage condition. Consider a model with two goods, X and Y, and two countries, H and F. The price of Y is normalized to 1. The price of X is p_H in country H and p_F in country F. A representative consumer in market H has a utility function $U(X,Y)$ and income I. Good X is subject to different standards in country H and F and importing X from country F to country H is associated with a variable arbitrage cost t and a fixed arbitrage cost T.

The arbitrage condition for country H can be stated in the following way. The utility received from buying the local-standard version of X at local prices must be at least as high as the utility gained from buying the foreign-standard version at foreign prices, including all arbitrage costs associated with the different standards. More formally, the individual arbitrage condition for good X in country H is

$$v(p_H, I) \geq v(p_F + t, I - T)$$

where $v(p, I)$ is the indirect utility function for the representative consumer in market H. For instance, consider a CES utility function

$$U(X,Y) = (X^b + Y^b)^{1/b}, \sigma \equiv 1/(1-b).$$

Maximizing utility with respect to income gives the indirect utility function

$$v(p,I) = I\left[\frac{(1+p^{\sigma-1})^{\frac{1}{\sigma-1}}}{p}\right].$$

We now study the effects of fixed and variable costs associated with different national standards. Evaluating the general indirect utility function at the fixed cost T gives an arbitrage condition for country H at pure fixed costs.

Figure 4.19 illustrates the arbitrage condition graphically. The domestic consumption bundle is A at price p_H and income I in country H. For some fixed cost T and price p_F in country F a representative consumer in country H is indifferent between A and buying the goods in F at the foreign consumption bundle A'. It follows that our arbitrage condition is satisfied for any price at least as high as p_F in country *F*. Moreover, arbitrage may be unprofitable despite a very high price in the domestic market. At price p_F in country *F* and a fixed cost higher than T_{lim} arbitrage is not profitable in market *H* at any domestic price.

Evaluating the arbitrage condition for variable cost *t* is straightforward. It implies that the conventional arbitrage condition must hold. The consumer would thus be indifferent between p_H and p_F at income *I* if and only if the variable transportation cost resulted in an optimal choice at point A, both at domestic and foreign prices.

The previous results may be used to study the welfare effects of a consumption distortion caused by different national standards in an international monopoly. If arbitrage were permitted it would impose a restriction on the international price differential but no arbitrage would actually occur in equilibrium. Assume that sector Y is perfectly competitive with marginal cost equal to one. Good *X* is produced at marginal cost mc_j in market $j=H, F$. The quantity $X_j\ (p_j\ )$ is the consumption of *X* in market *j*. Allow a single manufacturing firm in sector *X* to sell the product in both markets. The costs of different standards apply to individual arbitrage only and the manufacturing firm can distribute the two different standard versions of *X* without additional costs. Since trade is costless for the manufacturing firm and the marginal cost is

FIGURE 4.19. Individual arbitrage conditions

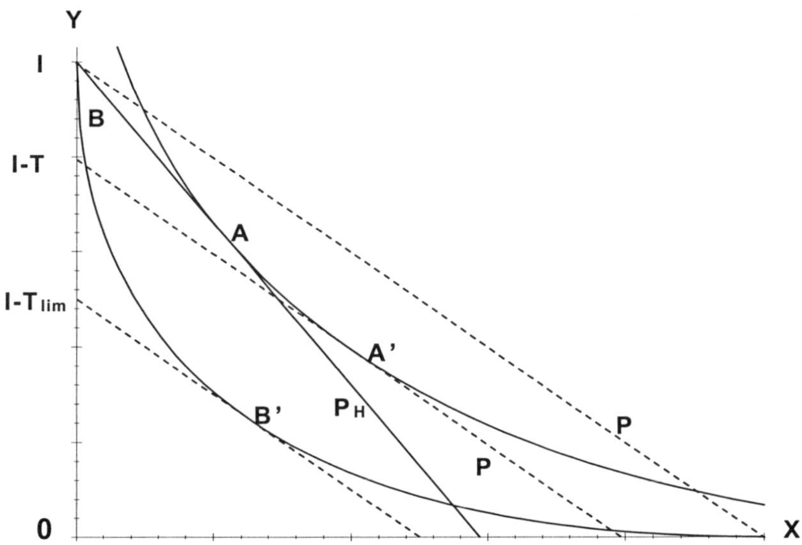

fixed, production can be located in country H or F without any implications for the equilibrium.

The monopolist in sector X would set prices p_H and p_F to maximize

$$\sum_{j=H,F} X_j(p_j)(p_j - mc)$$

subject to the arbitrage condition.

Figure 4.20 presents simulations of this partial-equilibrium model with fixed and variable arbitrage costs of different national standards. To capture different standards, we assume that the elasticity of substitution is in market H and in market F. However, the marginal cost is equal in both markets, i.e. $mc_H = mc_F = 0.50$.

By our computations, in the perfectly segmented equilibrium with prohibitively costly arbitrage the monopolist would set prices $p_H = 1.366$ and $p_F = 1$. Figure 4.20 plots the price differential $p_H - p_F$ for different variable and fixed arbitrage costs when individual arbitrage is permitted. For $T=0$, the price differential would increases linearly in the variable arbitrage cost up to $t=0.366$, where arbitrage would become unprofitable at segmented prices. Even low values of the fixed arbitrage cost T, on the other hand, would make arbitrage unprofitable. For any $T > 0.14$, prices would remain fully segmented. Even a

FIGURE 4.20. Price differential in an international monopoly

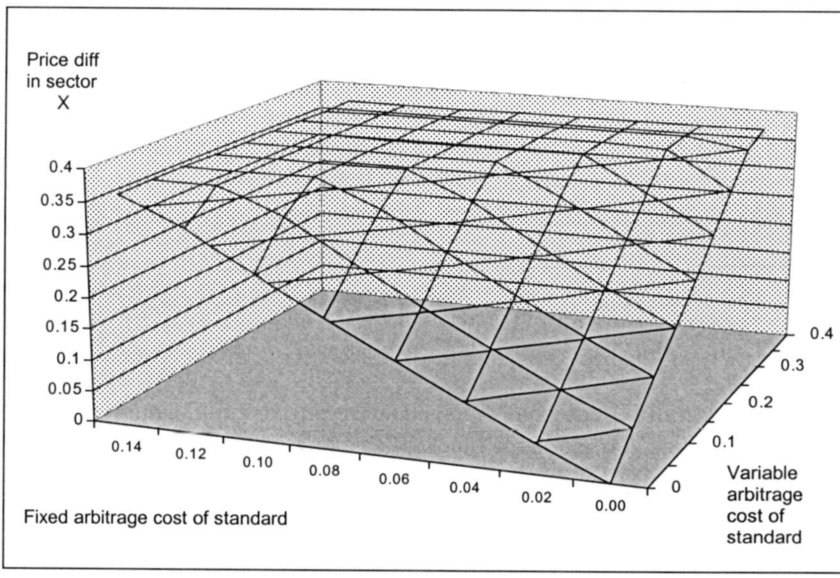

small fixed arbitrage cost of different national standards could, therefore, be prohibitive and support international market segmentation.

We also compute changes in welfare, though we do not show them. The simulation shows that total welfare including monopoly profits would rise by perhaps two percent in the integrated equilibrium, compared to its level in the segmented equilibrium. The main reason for the increase in total welfare is that equalized prices in the two countries would result in a reduction of the international consumption distortion caused by different local prices. However, it is also interesting to note that the benefits would be asymmetrically distributed between countries *H* and *F*. In brief, when markets are integrated consumer welfare would rise in *H*, where the price declines, and fall in *F*, where the price rises. Indeed, consumer welfare in country *F* would fall by approximately six percent as the arbitrage costs of different standards are reduced. Correspondingly, the consumer welfare in country *H* would increase by more than nine percent. Monopoly profits would be reduced as well.

The welfare effects in previously low-price countries can, however, be quite different when the cost of the different standards apply not only to the arbitrage activity but also to the output decision of the monopolist. If different national standards raised the real production and distribution cost for the manufacturing firm, its optimal output decision would change.

One important difference between a fixed and variable trade cost of different national standards is that the latter (but not the former) could change the marginal incentives of the monopolist. This effect can be studied in our model. Let the variable cost of different standards apply to both arbitrage and output by the monopolist. We first consider production of good *X* in country *F*, with exports to country *H*, after which we reverse these decisions. We assume that the arbitrage cost and the real trade cost of different standards are the same for individual consumers and the manufacturing firm. The marginal cost in market j is $t + mc_j$ when the monopolist exports the product to market j (assuming a symmetric cost from *H* to *F* and *F* to *H*). It is assumed that the individual arbitrage condition holds in equilibrium. In other words, arbitrage is a constraint on the international price differential in sector *X*, but no arbitrage occurs in equilibrium.

Figure 4.21 shows simulated relative price levels in both countries when the monopolist exports from market *F* to *H*. If the variable trade cost were increase, the higher marginal cost in market *H* would result in a higher optimal price there. The higher price in market *H* would induce a higher (not lower) price in market *F* due to arbitrage at low variable costs. However, for high variable costs the revenues earned in market *H* would be a small share of total profits of the monopolist and the arbitrage condition would have less impact on the price in market *F*. Eventually that price would fall.

FIGURE 4.21. Prices in a monopoly with trade and arbitrage costs (production in F, exports to H)

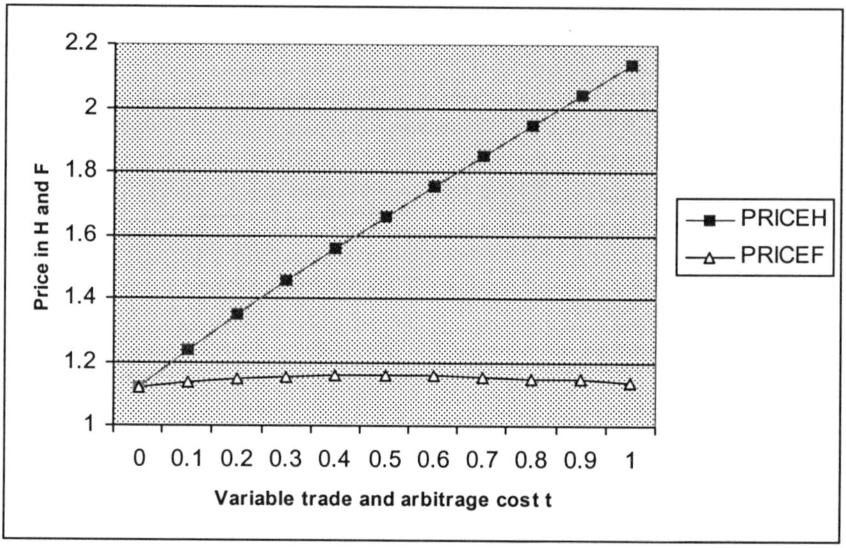

FIGURE 4.22. Prices in a monopoly with trade and arbitrage costs (production in H, exports to F)

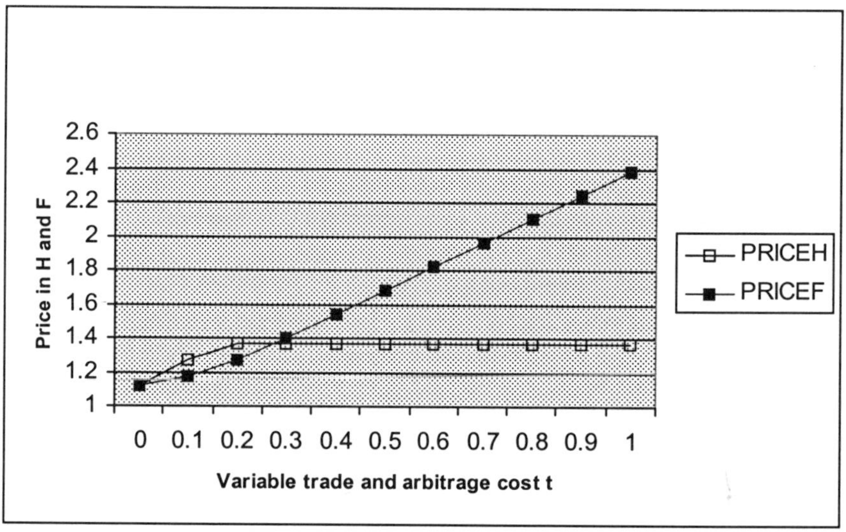

Welfare impacts in market F (not shown) would be a non-monotonic function of variable arbitrage costs. In particular, consumers in market F would prefer to see no difference in national standards, and therefore no variable cost, to standards that result in a moderate variable cost. For consumers in market H welfare gains from the integrated equilibrium would dominate any other equilibrium. Next, Figure 4.22 shows prices in market H and F when the monopolist exports from H to F. A higher marginal cost in market F (including the trade cost) would support a higher price there. For low variable costs the arbitrage condition would hold with equality but the effect on the price in market H would be weak. For high variable costs the arbitrage condition would not bind and the price in market F would exceed that in market H. Again, consumers in H would prefer the integrated equilibrium. Moreover, in this case consumers in F would prefer the integrated equilibrium to any difference in national standards.

Standards and Parallel Trade

More important than individual arbitrage is commercial arbitrage, where the activity is undertaken by international trading firms. Commercial arbitrage in products protected by intellectual property rights is referred to as "parallel imports" or "gray market imports" when the trade occurs without the intellectual property right holder's consent. Although commercial arbitrage is a wider term than parallel imports we will use the terms synonymously in the following analysis.

The most important reason that commercial arbitrage bears an advantage over individual arbitrage is that there may be increasing returns to scale in the trade activity. A fixed cost associated with different national standards could make individual arbitrage prohibitively costly, but still feasible for commercial arbitrage firms.

We extend the model from the previous subsection to allow for commercial arbitrage. The model still has two goods and two countries. We still assume that H is the high-price country and F is the low-price country in a segmented equilibrium. The costs of different standards apply to arbitrage activities but do not apply to production and distribution by the manufacturing firm. A single monopolist produces X for markets H and F at marginal cost mc_j in market $j=H, F$. Without loss of generality we assume that production is located in H, while the monopolist chooses outputs for both markets. We permit n arbitrage firms non-cooperatively to ship a quantity x^a from the high-price market to the low-price market. In its shipment decision each arbitrage firm takes the other arbitrage firms' quantities as given.

The condition under which commercial arbitrage into country H exists may be stated as follows. The price differential between market H and market F must be less than or equal to the average cost for an arbitrage firm to import the product from market F. More explicitly, the commercial arbitrage condition for good X in country H is

$$| p_H - p_F | \leq \frac{T}{x^a} + t$$

where x^a is the quantity shipped from market F to market H by the arbitrage firm. The variable T is the fixed arbitrage cost and t is the variable arbitrage cost. This condition permits free entry into the arbitrage activity. Arbitrage firms enter until there is no excess profit in equilibrium.

In the following analysis we focus on a pure fixed arbitrage cost of different national standards and no variable arbitrage cost, setting $t=0$. Assuming that all arbitrage firms are identical the total arbitrage quantity would be

$$X^a = \frac{(p_H - p_F)n}{-dp_H / dX - dp_F / dX}$$

This relationship holds subject to the condition $X_j(p_j) - X^a_j(p_H, p_F) \geq 0$, which follows from the n first order conditions of the arbitrage firms. Let the exported quantity from F be denoted $X^a_F = -X^a$ and the identical imported quantity in H be denoted $X^a_H = X^a$. The monopolist in sector X sets prices p_H and p_F to maximize

$$\sum_{j=H,F} (X_j(p_j) - X^a_j(p_H, p_F))(p_j - mc).$$

Again, we assume that the elasticity of substitution is $\sigma_H = 2$ in market H and $\sigma_F = 3$ in market F and that the marginal cost is equal both markets with $mc_H = mc_F = 0.50$. The fixed arbitrage cost T per firm is defined in per-capita terms. Thus, $T = TF/S$, where TF is the fixed arbitrage cost of the standard and S is the number of consumers.

Profit-maximizing prices in the segmented equilibrium without arbitrage are $p_H = 1.366$ and $p_F = 1$. In Figure 4.23 we see how prices would change when more arbitrage firms enter the market. The price in H would fall sharply and the price in F would rise sharply when the first arbitrage firm enters the market. The prices in both markets would converge to $p_H = p_F = 1.117$ as the number of arbitrage firms goes to infinity. This outcome corresponds to the inte-

FIGURE 4.23. Standards and parallel imports

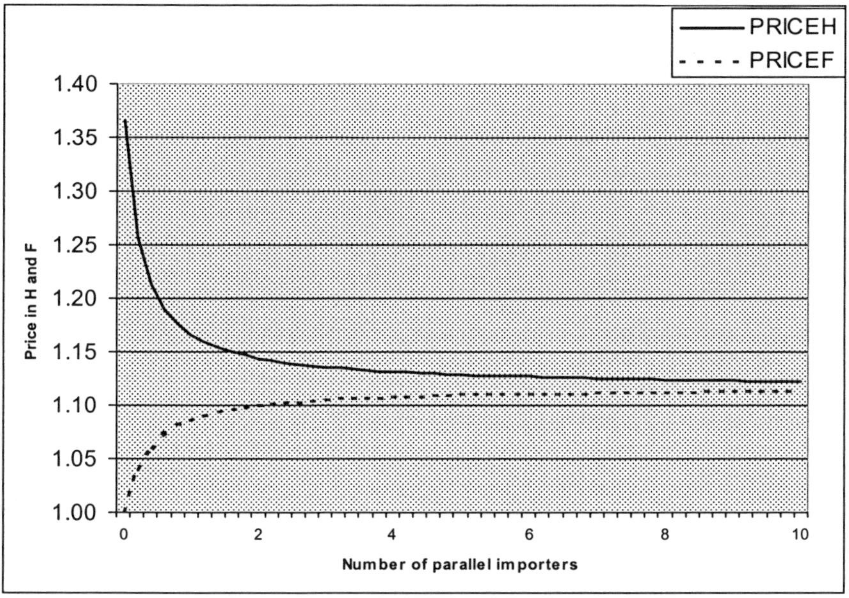

FIGURE 4.24. Standards and parallel imports

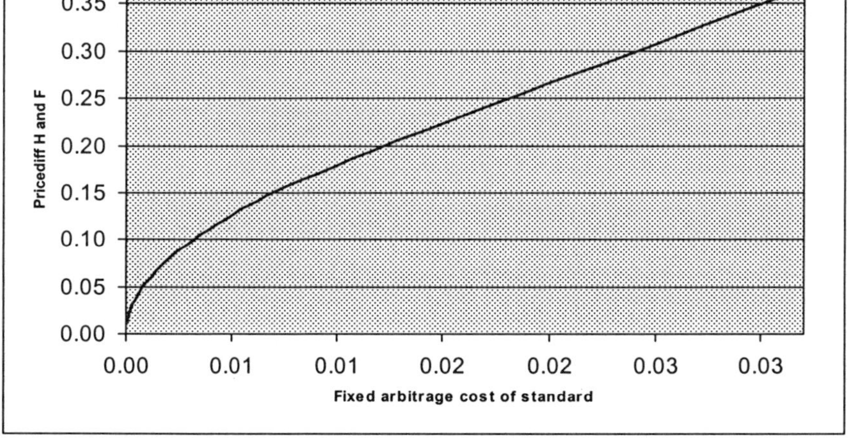

grated equilibrium price with costless arbitrage. Figure 4.24 illustrates how the price differential between *H* and *F* would depend on the fixed arbitrage cost of the standard. The fixed cost would determine the number of arbitrage firms. For a high fixed arbitrage cost, fewer arbitrage firms would find it profitable to enter. Consequently, the price differential would rise with the fixed arbitrage cost. The monopolist's profits would go up as costs increase.

Total welfare effects, which we do not graph here, would depend also on the size of the fixed arbitrage cost. Welfare, including monopoly profits, would be approximately two percent higher in the integrated equilibrium than under market segmentation. Total welfare would decline as the fixed cost increases. The main reason is that a larger price differential between markets *H* and *F* would generate an international consumption distortion like that in the previous section.

Standards and Marketing New Products

Different national standards may facilitate introduction of products into markets with large variation in preferences or income. More specifically, different local standards may promote investment in marketing and (partial) entry of products in low-income markets. A monopolist may find it profitable to introduce a new product in a low-income country as long as the supply in this market does not spill over to a high-income market. However, the introduction might be unprofitable if the monopolist is obliged to charge the same price to consumers in both high- and low-income markets. Under such conditions different national standards can be used to prevent arbitrage and facilitate marketing of new products in low-income, or low-valuation, markets.

The interaction between arbitrage and entry can be modeled as a discrete choice between entering and not entering a market. It can alternatively be modeled as a problem of partial entry, reflecting an investment in marketing. We modify our international monopoly model along the lines of this latter modeling approach. The monopolist incurs a cost $c(\theta_j)$ to inform a fraction θ_j of the consumers in market j about product X, where $c'(\theta_j)>0$, $c''(\theta_j)<0$. The resulting demand for product X would be

$$D_j(\theta_j, p_j) = \theta_j X_j(p_j)$$

and the monopolist would maximize

$$\sum_{j=H,F} D_j(p_j)(p_j - mc) - c(\theta_j).$$

In the segmented equilibrium, price in each market would be independent of the level of information available in the other. The first order condition with respect to θ_j is

$$\pi_j(p_j^*) - c'(\theta_j) = 0$$

where $\pi_j(p^*_j)$ is per-capita revenue in market j from informed consumers. Segmented prices would maximize $\pi_j(p^*_j)$ and information would be maximized at these prices. A consequence of this result is that arbitrage, generating a reduced price differential, would reduce information in both markets. In other words, different national standards and market segmentation can promote entry and investment across different markets.

Figures 4.25 and 4.26 illustrate simulation results for this model. We simplify the demand in market j to be $D_j(\theta_j, p_j) = \theta_j(a_j - p_j)$, where $a_H = 3$, $a_F = 2$. In the segmented equilibrium without arbitrage the optimal prices would be $p_H = 1.5$ and $p_F = 1$. For any variable arbitrage cost $t > 0.5$ segmented prices remain even in the presence of potential arbitrage. The optimal levels of information in the segmented equilibrium are $\theta_H = 0.562$, $\theta_F = 0.25$.

The effect of arbitrage on the information in market *H* and market *F* is illustrated in Figure 4.25. Compared to the segmented equilibrium the information in market *F* would be 13 percent lower in the equilibrium without different national standards (i.e. $t=0$). Correspondingly, the information in market *H* would be approximately one percent lower (right-hand scale). In other words, the monopolist would find it less profitable to invest in marketing in the low-price country when it is forced to charge the same price in both markets.

Figure 4.26 shows the effect on total welfare. We compare total welfare when the information in both markets is fixed by market segmentation with the welfare at endogenous information levels. Total welfare, including monopoly profits, is measured as an index equal to 100 in the segmented equilibrium. The result shows that the negative effect of less provision of information at low variable costs would dominate the positive effect of the reduced international price differential. Welfare in the integrated equilibrium would be 0.3 percent lower compared to the perfectly segmented equilibrium. Welfare would be 0.7 percent higher at intermediate variable arbitrage costs, due to a relatively small negative effect on information but moderate positive effect on the price differential.

FIGURE 4.25. Standards and information in an international monopoly

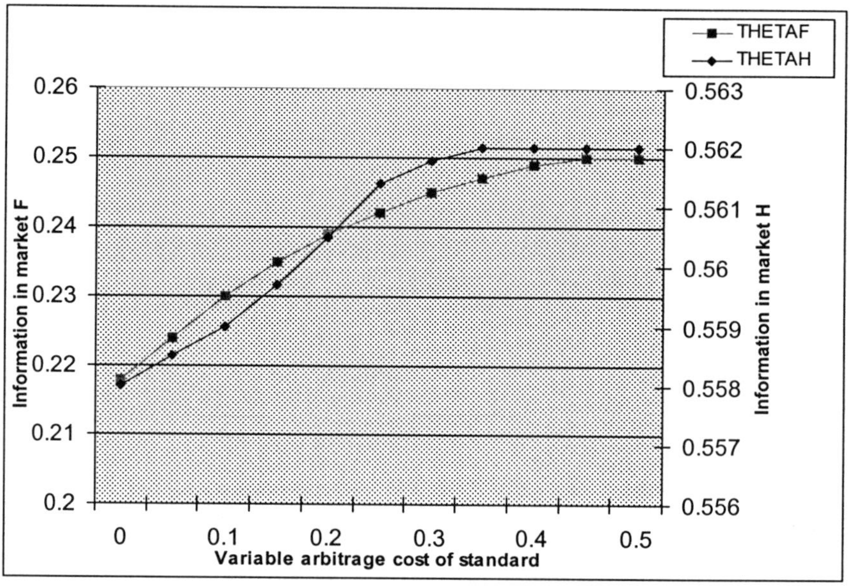

FIGURE 4.26. Welfare with endogenous information

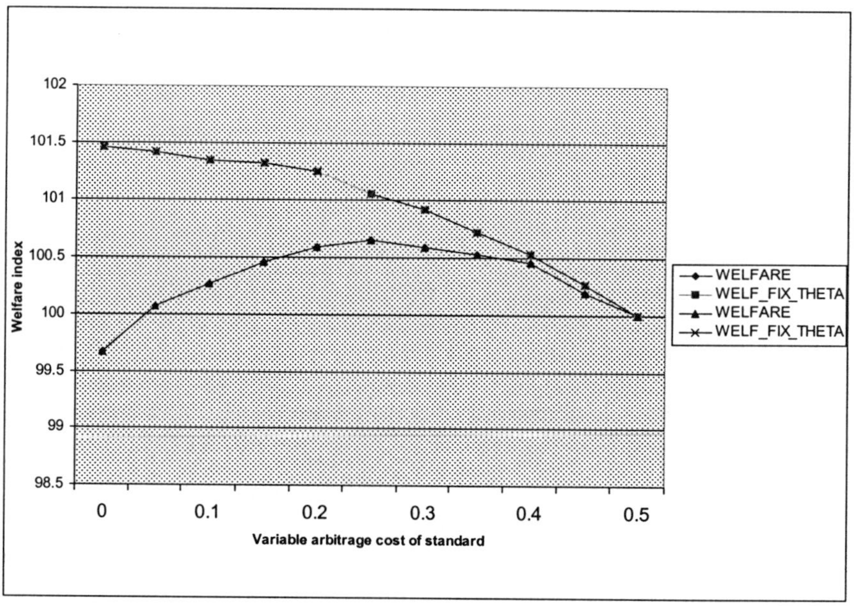

Standards and Production Efficiency under Increasing Returns to Scale

Thus far we have considered optimal pricing by a monopoly (similar results should hold for any fixed number of manufacturing firms) and we have concluded that different national standards can result in international consumption distortions but also enhance entry of products in low-income markets. Another important effect of standards, however, is their effect on production efficiency and the number of firms in equilibrium. We previously illustrated that reduced arbitrage costs against different standards could lead to reduced international price differentials in imperfectly competitive industries. Consequently, market integration could reduce the profits of firms, which could lead to exit and more efficient production. This effect may be illustrated in a simple model.

Consider again a model with two goods and two countries. The X sector is imperfectly competitive but entry and exit are free, with n firms incurring a fixed cost C and non-cooperatively choosing quantities for both markets. The marginal cost of production is assumed to be mc in both markets and for all firms. There are no trade costs and the costs of different standards do not apply to these manufacturing firms.

First, we assume that different national standards make arbitrage prohibitively costly and therefore perfectly segment markets. The first-order condition for profit-maximization seen by firm i in the X sector is

$$\frac{d\, p_j(X_j)}{d\, x_{ji}}\, x_{ji} - (p_j(X_j) - mc) = 0$$

for markets $j=F,H$, where x_{ji} is the quantity chosen by firm i in market j. The number of firms is determined by the free-entry condition

$$\sum_{j=H,F} (p_j(n\, x_j^*) - mc)\, x_j^* - C \leq 0$$

where n is the complementary variable and x^*_J the non-cooperative equilibrium quantity with n firms.

With a common standard in both markets we set markets to be perfectly integrated. In this case the first-order condition is reduced to

$$\frac{dp(X)}{dx_i}\, x_i - (p(X) - mc) = 0$$

where $p(X)$ is the inverse demand function for the integrated market and x_i is the quantity chosen by firm i for the integrated market. The number of firms is determined by the free-entry condition

$$(\, p(\, n\, x^*\,) - mc\,)\, x^* - C \leq 0$$

where n is the complementary variable.

We simplify demand in market j to be $D_j(p_j) = (\text{"}_j - p_j)$, where $\text{"}_H = 3$, $\text{"}_F = 2$. The fixed cost of manufacturing ranges from 0.0 to 1.0. In the equilibrium with a fixed cost equal to zero, price is set equal to marginal cost. Figure 4.27 illustrates impacts on equilibrium prices. Prices would increase as the fixed costs of production go up, as fewer and fewer firms would find it profitable to enter the market. In the segmented equilibrium, prices (PRICEHS and PRICEFS) would become increasingly more dispersed as the fixed cost increases. A smaller number of firms would give each remaining firm more market power and the scope for price differentiation would rise. The integrated price (PRICEHI) would be lower than the segmented price in market H but higher than the segmented price in market F. We also compute the integrated equilibrium price for the situation where the same number of firms exists as in the segmented equilibrium (PRICEC). This price would be lower than the integrated price as more firms would be active in equilibrium, but the difference would be small.

Figure 4.28 shows the effect on total welfare. Welfare is measured as an index set to 100 for the segmented equilibrium at every level of the fixed cost. Thus, it is reset to 100 in the segmented equilibrium for every level of C. Total welfare in the integrated equilibrium would be higher than the total welfare in the segmented equilibrium for any given fixed cost. The difference between total welfare in the segmented and integrated cases would rise with the level of fixed cost (WELF_T). Total welfare would be almost five percent higher in the integrated equilibrium than in the segmented equilibrium when the fixed cost is close to unity. The total welfare difference between the integrated and segmented equilibrium would be the sum of the reduced international consumption distortion and more efficient production when some firms exit the market. The positive effect of this reduced international consumption distortion is measured as the increase in welfare in an integrated equilibrium without exit (WELF_P). Thus, approximately 2/3 of the welfare gain may be attributed to a reduced international consumption distortion and 1/3 to more efficient production. Finally, we note that the benefits from market integration would not be distributed equally. Consumers in market F would be better off in the segmented equilibrium and consumers in market H would be better off in the integrated equilibrium.

FIGURE 4.27. Arbitrage and production efficiency effects on prices

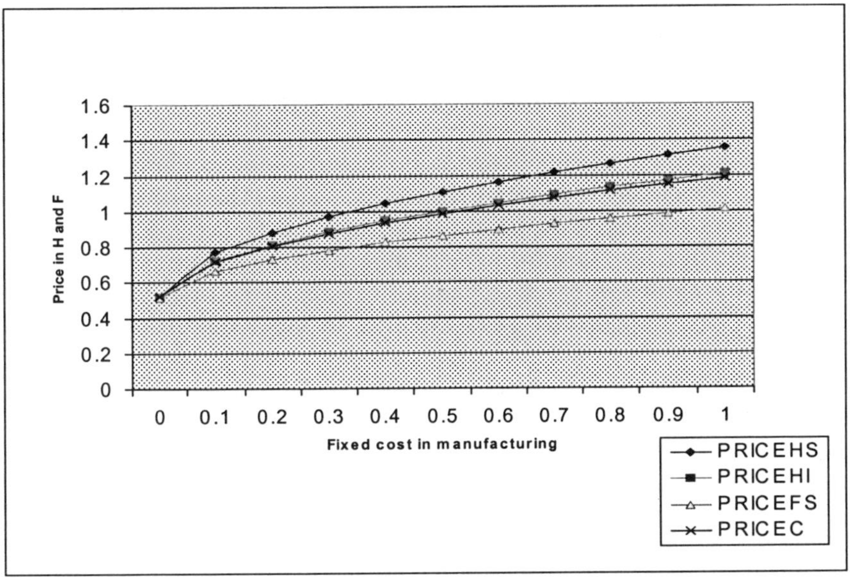

FIGURE 4.28. Arbitrage and production efficiency
- relative welfare compared to segmented equilibrium

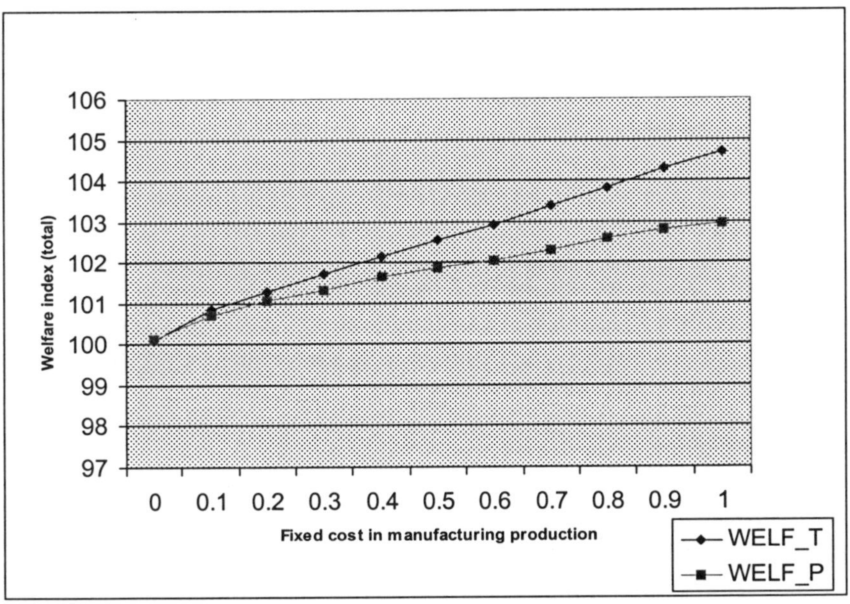

Summary

The purpose of this chapter was to consider approaches to formally modeling standards and technical regulations governing trade. An important criterion for the experiments was that they be potentially implemented in applied general-equilibrium models with real data. Our simplest approach was to suppose that when they differ between countries, standards constitute a real trade cost for exports trying to penetrate the foreign market. Our results in this case indicated that incompatible standards are particularly harmful for smaller, poorer countries, which cannot win a trade war fought over standards.

We considered this situation to be too simple in many cases. In particular, standards may prove beneficial to exporters by certifying their products as safe, healthy, or compatible with local complementary goods. In Section 3, we therefore added to our analysis by showing alternative ways to model an increased willingness to pay on the part of local consumers for imported products meeting the domestic standard. We emphasized that this must be done with care, particularly because different methods of modeling a change in willingness to pay would leas to different welfare effects.

In Section 4 we noted that standards might operate more on fixed cost than on variable cost. If this cost were a quasi-public good across firms, there could be multiple equilibria. In one case, firms would export while in another they would not export. We illustrate that the welfare differences between these two outcomes may be large, supporting an important coordination role for public policy.

In Section 5 we analyzed how standards might alter the substitutability or complementarity between local and foreign products. Again we emphasized that this choice must be modeled carefully so as not to produce misleading welfare results. Our examples indicated that consumers and imperfectly competitive firms would have opposing interests when standards increase substitutability, but have reinforcing interests when standards increase complementarity. Finally, in Section 6 we considered how standards could alter the ability of consumers and independent firms to conduct arbitrage and parallel trade. We assumed that the original manufacturer had no costs in meeting different standards (and indeed sometimes created them precisely to defeat arbitrage) while other firms and consumers did have such costs. Under plausible assumptions our model predicted that the manufacturer might set a lower price in a foreign (e.g., developing) country than in its home country if it could segment the two markets via different standards.

The ability of independent firms to arbitrage against such price differences could increase total world welfare but worsen welfare in the developing country. This outcome would emerge because the manufacturer would be obliged to raise prices there to prevent arbitrage or would engage in less informative advertising.

In all sections, we developed simple numerical models and solved them, thereby illustrating that our approaches are implementable. Significant obstacles to full implementation with real data remain, however, since the beneficial effects of standards, and indeed often the costs, are largely unobservable in published data.

NOTES

1. By identical relative technologies or productivities across sectors we mean that there is no Ricardian comparative advantage. One country could be uniformly more advanced in all sectors.

2. We have not shown aggregate welfare effects in the two countries to avoid clutter. The large country would not benefit in aggregate from its own standard (Figure 4.1). Thus, a 25 percent rise in profits cannot compensate for a 5% fall in consumer real income. In the simulations, neither country would benefit in the aggregate, which would be true also for the small-country standard.

3. This general-equilibrium effect is missed in partial-equilibrium models. As output for supply to the small market falls for the large-country firms, factor prices would fall for the X industry in country h. Thus marginal revenue (initially unchanged) would exceed marginal cost for sales to the h market and supply would increase. This would not happen in familiar strategic trade-policy models with constant costs.

CHAPTER 5

Quantifying the Trade Impact of Compatibility Standards and Barriers: An Industrial Organization Perspective

Neil Gandal

Introduction

The goal of this chapter is to help set a research agenda for examining the effect of compatibility barriers to trade. The chapter explores the relationship between compatibility standards and international trade flows from an industrial organization perspective. I provide a brief survey of the literature on the economics of compatibility and standardization, as well as a brief survey of the literature on how standards affect international trade. I then discuss several industries that provide "natural experiments" to measure the effect of compatibility standards on international trade. In doing so, the chapter brings together two separate literatures:

The economics of compatibility and standardization: Despite the fact that many industries characterized by "network effects" (personal computers, telecommunications, consumer electronics products) are global, this literature has almost exclusively focused on closed economies. The analysis of compatibility standards differs between closed-economy and open-economy contexts for several reasons. One difference (which also applies to a setting where there are no network effects) is that in the case of open-economies, foreign firms' profits do not contribute to the domestic country's welfare. A perhaps more important difference is that the analysis of closed economies ignores any gains that might come from international coordination of standards. When there are network effects, the benefits from standardization increase in the size of the network, regardless of whether the consumers are foreign or domestic.

The literature on barriers to international trade: This literature has focused for the most part on the strategic effects and welfare consequences of "traditional" trade barriers such as tariffs, quotas, and Voluntary Export Restraints (VERs). To the extent that the effects of strategic standardization policy have been examined in the trade literature, they have been considered primarily in the context of minimum quality (vertical) standards, rather than compatibility standards.

Section 2 discusses the standard setting process. In section 3, I survey the economics of compatibility and standardization, while section 4 provides a survey of the trade literature on the effect of standards. section 5 discusses empirical (trade) implications of standardization policies. In section 6, I discuss two industries that provide "natural experiments" to measure the effect of compatibility standards on international trade. Section 7 discusses two additional industries where compatibility seems important, but where it may be difficult to empirically test for the effects of compatibility on trade. Section 8 provides brief conclusions.

Setting Standards

Broadly speaking, there are three ways that standards get set:

I. De facto standards, i.e., standards set primarily by market. These standards are often proprietary:

- The PC operating system industry provides an example. Due to a bandwagon effect, Microsoft has succeeded in setting standards in the PC operating systems industry. In 1992, it was estimated that seventy-two percent of all new personal computers sold throughout the world were shipped with the MS-DOS operating system.[1]

- In the case of video cassette recorders (VCRs), two incompatible proprietary standards competed. Sony, the initial leader, did not offer attractive licenses for its Betamax technology to other firms that wished to make VCRs, while JVC, a relatively unknown firm did just that. The Betamax technology was apparently - "on its own" - as good as the competing incompatible VHS technology.[2] Nonetheless, by 1981, VHS held a 66-percent share of the VCR installed base.[3] When pre-recorded video cassettes became important in the early 1980s, rental stores preferred to carry VHS tapes because of their compatible installed-base advantage. The dearth of Betamax tapes "tipped" the market to VHS, which became the de facto standard in 1988. See Grindley (1995) for more details.

II. Voluntary industry agreements, where standards are often jointly developed. These standards are typically open standards, they are not proprietary.

- Compact-disc (CD) technology was developed by Philips in 1979 and introduced to the United States by Philips and Sony in 1983. In order to encourage adoption and to sell their software products (Philips owned Polygram Records and Sony owned CBS Records of Japan), Sony and Philips licensed their technology quite liberally. McGahan (1991) and Grindley and McBryde (1992) note that by 1981, more than 30 firms had signed licensing agreements to use the Philips technology and that other firms had withdrawn competing prototypes. Consequently, by the early 80's, CD-players had become a fairly standardized product produced by many firms.

- The DVD (digital video disc) industry provides an example of a jointly developed standard. Throughout the 1990s, video hardware and software manufacturers sought a digital format to replace videocassettes. In order to avoid another Beta/VHS format war, hardware manufacturers led by Sony, Toshiba, and Panasonic, and movie studios, led by Warner Bros. and Columbia Pictures (a division of Sony), worked together to establish a single standard, resulting in a non-proprietary DVD standard.

III. Standards imposed by National Standards Bodies (NSBs), or agreed upon by regional or international standards development organizations (SDOs):

- NSBs such as the U.S. Federal Communications Commission (FCC) coordinate national standardization activity for all industries and use of their standards is mandatory. SDOs, on the other hand, are specialized along industrial organization lines (i.e., by industry). Examples of SDOs include the International Telecommunications Union (ITU), the oldest international standards body in the world, and the International Electrotechnical Commission (IEC). In the case of telecommunications, governments typically feel compelled to require compatibility and interoperability standards. Given the importance of compatibility among international phone networks, the standards are set by the ITU are done so by international consensus. Similarly, IEC standards are typically set by international consensus.

- David and Shurmer (1996) provide qualitative evidence that there has been an increase in international standards production. The evidence includes a doubling of number of IEC standards between 1986-1990. They also note that there has been a shift from national standards to regional/international standards. The shift towards regional/international standards is due, in large part, to advances in information & communications technologies (ICTs) and the importance of technical compatibility within products (such as computer operating systems & applications software and consumer electronics products) that employ these new technologies. According to the authors, "the ICT impact upon the standardization regime have been profound (p. 797).[4]"

Two important welfare implications of network effects are that: (i) markets may fail to achieve standardization when it is socially desirable, and (ii) even if the market or regulators do agree on a standard, the chosen standard may be inferior. Both of these "failures" are due to coordination problems.

The Economics of Compatibility and Standardization: Network Effects

A network effect exists when the value that consumers place on a particular product increases as the total number of consumers who purchase identical or compatible goods increases. In the case of an **actual (or physical) network**, such as the telephone network, the value of the network depends on the total number of subscribers who have access to the network.

In the case of virtual networks that are not linked physically the network effect arises from **positive feedback from complementary goods**. Examples of virtual networks in which the value of the "base" product increases as the variety of complementary products increases include computer operating systems, videocassette recorders (VCRs), compact disc players (CD-players), and Digital Versatile Disc players (DVD-players). In the case of computer operating systems, the complementary goods are the applications software programs, while in the case of VCRs, the complementary goods are the VCR cassettes or tapes; similarly in the case of CD-players, the complementary goods are the compact discs, while in the case of DVD-players, the complementary products are the DVD-discs.

The positive feedback mechanism works as follows: the value of the base product is enhanced as the variety of (compatible) complementary products increases; hence consumers will be more likely to purchase a base product with many compatible complementary products. The variety of complementary products, in turn, will depend on the total number of consumers that purchase the base product. As the number of consumers that purchase the base product increases, there is a greater demand for compatible complementary products. This increases the profitability of supplying complementary products. Since there are typically fixed or sunk entry costs, production of the complementary products is characterized by increasing returns to scale. Hence more complementary products will be produced or developed for a base product with a large share of the market. This further enhances the value of the base product. Thus there is positive feedback in such a system: an increase in the sales of the base product leads to more compatible complementary products, which further increases the value and sales of the base product.

As Katz and Shapiro (1994) note, the positive feedback means that there is a "natural tendency towards de facto standardization" (p.105). They note that these system markets are often characterized by tipping: once a system has gained an initial lead, there is a snowball effect. One system ends up being the market standard with large amounts of compatible complementary products; the other system has a very small market share, if any at all. The value of the base product with little or no complementary software is essentially zero, since the base product itself provides little or no standalone benefits.

Theoretical Literature

The idea of network effects was first enunciated by Rohlfs (1974). In the mid-eighties Katz and Shapiro (1985), (1986) and Farrell and Saloner (1985) extended this idea to the oligopoly context, and examined the social and private incentives to achieve compatibility in a single product network. Matutes and Regibeau (1988) examined similar issues in a setting in which products consist of components that must be used in a fixed (one to one) proportion. Chou and Shy (1990) and Church and Gandal (1992) examined related issues in systems markets. The two critical welfare issues discussed above (suboptimal standardization & choice of an inferior standard) have been examined in these and other papers.

While the theoretical literature of the economics of compatibility and standardization is quite large, nearly all of the literature has examined closed economies. See David and Greenstein (1990), Gilbert (1992), and Matutes and Regibeau (1996) for more detailed surveys.

Empirical Literature

A small but growing literature has empirically (statistically) found evidence of virtual network effects. In a study of the mainframe computer market, Greenstein finds that other things being equal, a firm with an IBM 1400 was as likely as any other firm to purchase an IBM when making an additional purchase. On the other hand, he finds that a firm with an IBM 360 was more likely to purchase an IBM than a firm that did not own an IBM 360. Software for the IBM 1400 could not run on the following generations of IBM models (360, 370, 3000, and 4300), while software for the IBM 360 could run on the 370, 3000 and 4300. Hence, Greenstein's results can be interpreted as a demand for compatibility.

Gandal (1994) estimates quality adjusted price equations for computer spreadsheet programs from the 1986-1991 period and then uses the analysis to empirically test whether network effects exist in this industry. The study finds that, other things being equal, consumers were willing to pay a significant premium for spreadsheets that were compatible with the LOTUS file compatibility standard.

Saloner and Shepard (1995) test for the existence of network effects in the adoption of automatic teller machines (ATM's). In particular, they test whether banks with a larger expected number of ATM locations will adopt the ATM technology sooner. Since expected network size is not an observable variable, they used the number of branches as a proxy. Their results indicate the presence of network effects.

Gandal, Greenstein, and Salant (1999) provide empirical evidence that the personal computer operating system market exhibits (positive feedback) virtual network effects. They examined the early micro or personal computer market (the 1978-1986 period) in which CP/M, the early de facto standard operating system, was subsequently replaced by the DOS operating system. The paper found statistically significant two-way feedback between advertisements for the operating system and advertisements for applications software for both CP/M (for the 1978-1986 period) and DOS (for the 1981-1986 period).

Gandal, Kende, and Rob (2000) examine the diffusion of the CD player/disks "hardware/software system." For such systems there is interdependence between the hardware-adoption decisions of consumers and the supply decisions of software manufacturers. Hence there can be bottlenecks to the diffusion of the system. Using a structural model, they estimate the (direct) elasticity of adoption with respect to CD-player prices and (the cross) elasticity with respect to the variety of CD-titles. The results show that the cross elasticity is significant.[5]

Despite the fact that many goods characterized by network effects are traded internationally, the literature on network effects has almost exclusively focused on closed economies. Indeed, to the best of my knowledge, there is no empirical work on the relationship between compatibility standards and international trade. This is most likely due to the fact that relevant theoretical literatures (the industrial organization literature on network effects & the trade literature on the effect of standards) have not addressed the international trade issues. Although systematic empirical work in this field is demanding, the data to examine the relationship between compatibility standards and international trade are available.

Trade Literature on the Effect of Standards

Literature on standardization policy and international trade has primarily examined the effects of imposing minimum quality (vertical) standards. Recent contributions on the effect of imposing minimum quality standards include Barrett (1994), Boom (1995) and Lutz (1996). The latter two authors employ a model of vertical product differentiation to examine the effect of minimum quality standards. Barrett (1994) examines the incentives for governments to impose environmental protection standards on industries that compete in oligopolistic international markets. Casella (1996) reviews the literature on minimum quality (vertical) standards.

The small theoretical literature on compatibility standards and international trade includes Kende (1991a) and Shy (1991), who respectively consider the effect of compatible international standards on licensing and on the incentives for conducting R&D. Kende (1991b) shows that standardization increases a domestic firm's profits internationally because of the increase in the product's gross utility. Jensen and Thursby (1996) examine strategic incentives for setting standards. In their paper, governments set standards in order to improve the chances of domestic firms in R&D competition.

None of the above papers examine governments' strategic standardization policies with network effects in an international environment. The extension of the industrial organization literature on compatibility and standardization to an international environment means that there are domestic and foreign firms. The key consideration is that foreign firms' profits do not contribute to domestic country's welfare. Hence, optimal standardization policies may be different between open and closed markets. Sykes (1995) verbally discusses (in a trade context) the "standardization" tradeoff between network effects and increased costs for foreign producers.

Gandal and Shy (2000) formally examine strategic aspects of governmental standardization policy and the welfare implications when products and standards are horizontally differentiated. They analyze governments' incentives to recognize foreign standards when there are potentially both network effects and conversion costs. Network effects provide consumption benefits, while requiring foreign firms to comply with domestic standards may raise the costs of foreign producers.[6]

They employ a three-country, three-variety, single-industry model. Each of three horizontally differentiated brands is produced in one of the three countries. Using an industrial organization, intra-industry trade framework, they examine a setting in which countries can form standardization unions. The interaction takes place in stages. In the first stage, each government decides whether or not to recognize foreign standards. In the second stage, each firm sets profit maximizing prices in each country and consumers make purchases.

When conversion costs are large, relative to network effects, two countries can increase their welfare by forming a standardization union which recognizes the standards of both member countries, but does not recognize the standard of the third (nonmember) country. When network effects are large, relative to conversion costs, all countries will mutually recognize all standards, that is, they have no incentive to form exclusionary standardization unions.

The results as to when countries form or do not form exclusionary standardization unions are somewhat specific to the setting considered. The main message of the paper is that in a setting in which network effects dominate conversion costs, there is more international trade than in the setting where network effects are small relative to conversion costs. This conclusion is likely quite robust.

Empirical Implications of the Effect of Standards on International Trade

Modern international trade theory recognizes that most industries are characterized by product differentiation and economies of scale. These two effects typically lead to some type of imperfect competition. Using standard industrial organization models, it is straightforward to show that economies of scale and product differentiation give rise to intraindustry trade. Despite the fact that there is virtually no theoretical literature on the effect of compatibility standards on international trade, the industrial organization (imperfect competition) framework provides several testable predictions regarding intraindustry trade:

1. In an oligopoly setting with differentiated products, compatibility i.e., standardization should lead to an increase in product sales. This is because of the network effect, i.e., an increase in the number of consumers using compatible products increases the value of the product itself. This in turn leads to additional sales.

Thus when there is free entry into the provision of a standardized product, compatibility typically will lead to the entry of additional firms into the market as a response to the increase in market size. Since firms generally produce differentiated products, open standards should lead to increased international trade.

2. Firms in developing countries are more likely to enter industries where standardization has been achieved due to the larger market size. This is a corollary of the prediction discussed above. The OECD (1996) attributes the growth of developing countries exports in the consumer electronics industry, in part, to the fact that these countries are low-cost producers of **standard** products.[7]

3. The entry of additional firms means that compatibility should in general lead to increased price competition. This is a straightforward result from an imperfect competition, discrete choice model of product differentiation. The successful "cloning" of the IBM PC is an example of this effect. (Section 7.1.)

It is important to point out, however, that under settings in which firms sell systems composed of components, incompatibility may actually lead to lower prices than compatibility.[8] Given that the number of firms is fixed in these settings, this effect is likely temporary.

Industries to Study

Ideally, one would like to control for all other variables and examine the trade implications of compatible vs. incompatible standards. Since such controlled settings do not exist, the next best alternative is to examine industries in which the course of events was such that it is possible to isolate the effects of compatibility on trade. I refer to such settings as "natural experiments." In this section, I briefly describe two industries that provide natural experiments to measure the effect of compatibility standards on international trade. In both of these industries, it should be possible to collect the relevant data.

Television Industry

The National Television Standards Committee (NTSC) system was developed in the U.S. in 1954. The Sequential Couleur Avec Memoire (SECAM) system and the Phase Alternate Lines (PAL) system were developed in the early 1960s, SECAM in France and PAL in West Germany. All three standards are incompatible. The U.S. and Japan adopted NTSC, while the PAL system was adopted by most countries in Western Europe (except France). France and Eastern European countries adopted SECAM.

This fragmentation likely slowed the development of a global market for television receivers. Indeed, European governments in part adopted color television standards that were different from the U.S. standard in part to protect the interests of domestic firms. In order to sell television sets in France, foreign manufacturers had to adapt (convert) the receivers to the SECAM standard.[9] Burton and Saelens (1987) note that Japanese entry into the US occurred before Japanese entry into the European markets, in part, because both countries were on the NTSC standard. Table 5.1 shows that Japanese sales (Japanese imports and local Japanese production) in the U.S. took off in the mid 1970s. In 1974, Japanese color television receiver sales imports plus local Japanese production) accounted for 15.6 percent of color television sets sold in the U.S. By 1976, Japanese color television receiver sales accounted for 47.3 percent of color television sets sold in the U.S. Despite the decrease in Japanese imports, the increase in local production by Japanese subsidiaries insured that the Japanese manufacturers continued to account for a significant portion of television sales in the United States.

In 1981, sales by Japanese firms accounted for 43.5 percent of color television sales in the U.S. In contrast, sales by Japanese firms accounted for only 15.2 percent of color television sales in Western Europe in 1983.[10]

Burton and Saelens (1987) remark that "access to European markets was impeded by technical and licensing barriers," (p. 289) where the technical barrier refers to the incompatible standards and the licensing barrier refers to the fact that Telefunken, owner of the PAL technology, refused to provide Japanese firms with licenses until 1970. The 1988 world-wide installed base for the three competing systems were:

NTSC: (established 1954): 280 million (US 195 million, Japan 71 million, and Canada 14 million).

PAL (established 1967): 270 million (UK 30 million, Germany 23 million, Italy 22 million, Spain 13 million, plus all other European countries besides France).

SECAM (established 1967): 150 million (USSR 107 million, France 22 million, Eastern Europe 22 million).

Given the size of the U.S. market and the early entry of Japanese firms into the U.S. television market, it is likely that television prices (adjusted for quality) were higher in Europe (especially France) than in the U.S. Data should be available to test this hypothesis. Gordon (1990) provides several quality-adjusted price indices for television sets for the U.S. for the 1952-1986 period. As long as "compatible" data are available from a subset of European countries, it will be possible to empirically test this hypothesis.

TABLE 5.1. Television Sales (in U.S. Millions)

Year	Demand	Japanese Imports	Local Japanese Production
1974	8.02	1.00	0.25
1975	6.65	1.22	0.58
1976	7.89	2.96	0.77
1977	9.40	2.14	1.16
1978	10.50	1.54	1.77
1979	10.24	0.69	2.27
1980	10.90	0.57	3.21
1981	11.16	1.02	3.83

Source: Burton and Saelens (1987).

According to Burton and Saelens (1987), by the early 1980s, "the technical barrier created by competing transmission systems had virtually disappeared because of advances in microelectronics. (p. 289)" Hence, it may be possible to examine whether the elimination of the technical (compatibility) barriers facilitated entry of developing country firms into the television receiver industry. The data necessary to examine this issue include export sales data for television receivers by country and year, as well as total sales data by country and year for the main markets for developing country exports. Such data is almost certainly available for the U.S. and the U.K. and possibly other European countries. It will be important to control for the general trend of increased exports from developing countries over time by using overall export sales data for these countries.

Additionally, it would be interesting to examine how compatibility between the U.S. and Japanese systems affected Japanese entry into the U.S. It may not be possible to conduct this test because other likely explanations for early Japanese entry into the U.S. market.[11] Nevertheless, it will be worthwhile to measure trade flows in this industry.

Consumer Electronics: Compatible CD players vs. Incompatible VCRs

According to the OECD, worldwide exports of consumer electronics products increased from approximately $17 billion in 1982 to $40 billion in 1992. The growth rate in exports for consumer electronics has been much higher among developing countries than among developed countries over the 1980-1990 period. The OECD (1996) attributes this growth, in part, to these countries being low cost producers of standardized products.

While there was standardization within many consumer electronic product categories, in the case of videocassette recorders (VCRs), incompatible proprietary standards competed. In the case of CD players a single worldwide open standard was agreed upon before product introduction.

A testable empirical implication is that other things being equal, firms in developing countries entered the CD industry more than quickly than the VCR industry. This hypothesis could be tested by collecting overall market sales data and export data from developing countries for both the CD and VCR industries by year. In order to conduct this test, one would have to control for variables such as prices, Gross National Product, and the availability of complementary products.[12]

Additional Industries

In this section, I briefly discuss two industries where compatibility seems important, but where it may be difficult to empirically test for trade effects.

Personal Computers Operating on DOS/Windows Operating Systems

CP/M was the dominant open operating system during the 1976-1980 period.[13] It was used on more than 50 microcomputers including Commodore, Hewlett Packard, and Zenith. CP/M's market share was between 20-30 percent in 1980 (Gabel 1991).[14]

The introduction of the IBM PC occurred in late 1981. The IBM personal computer used a 16-bit chip from Intel and the DOS operating system. The operating system, developed under contract by Microsoft, was called MS-DOS. Under the contract with IBM, Microsoft had the copyright to MS-DOS and also had the right to sell it or variants to other producers of microcomputers. By 1984, IBM's share of the personal computer market reached 33 percent (Gabel, 1991). By this time, the IBM PC with the DOS operating system had supplanted the CP/M operating system as the industry standard.

Gabel (1991) notes that during the IBM PC's first few years, IBM was able to earn a "digital handshake" premium. Grindley (1995) notes that in 1983, the IBM PC was priced at approximately $4000, while a clone was priced at about $2000, or half that level (pp. 134-135). Hence the digital handshake premium was quite a significant proportion of the IBM price. Gabel (1991) notes that the IBM premium was approximately $1570 in 1985.

There was a significant premium because, the early "clones" (PC-like machines) were not perfectly compatible, that is, they were not able to run all of the applications software written for the IBM PC and MS-DOS operating systems (Gabel 1991). This was due to the one feature of the PC that was proprietary: The ROM BIOS (Read-Only Memory Basic Input/Output System.) Once competitors managed to reverse engineer the IBM proprietary ROM BIOS, clones that were fully compatible with the IBM platform became available and they captured a very large share of the PC market.[15] As a consequence, the digital handshake premium for IBM machines declined. "But as the clones established a reputation for perfect compatibility, the premium attached to the IBM brand name diminished" (Gabel 1991, p.26).

Since virtually all PCs used the DOS/Windows operating systems, they were all fully compatible. Did compatibility lead to large flows of international trade in the PC industry? Did standardization lead to the entry of firms from developing countries in the PC market? While there are no obvious empirical tests, it would be interesting to quantify trade flows in this industry over time.

Wireless Telecommunications

In the fast growing world on wireless communications, several proprietary incompatible technologies have been competing throughout the world. (The discussion in this section is based on Funk (1998) and Grindley, Salant and Waverman (1999).) The "first generation" of analog cellular standards came into being in the early 1980s. The dominant standard was the advanced Mobile Phone Service Standard (AMPS) which operated at 800MHz. AMPS had its origin in the U.S. in 1983, and was adopted by the U.K. in 1984. By 1992, AMPS had 17.6 Million subscribers. This grew to 75.0 Million in 1996, with more than half of the AMPS subscribers in the U.S.

In 1987, the operators from the major European countries agreed to deploy digital Group Speciale Mobile (GSM) at the same frequency in order to enable roaming, that is, in order to enable consumers to be reached outside of their home service area. The European Telecommunications Standards Institute (ETSI), which was formed in 1988, facilitated this process. GSM was introduced in 1992. By 1996, GSM had 32.5 Million subscribers. By 1999, it had 70 million subscribers and was being used in 70 networks.

No formal digital standard was set in the U.S. The Telecommunications Industry Association (TIA) recognized GSM, as well as two additional "second generation" digital standards that were developed in the U.S.: Code Division Multiple Access (CDMA) and Time Division Multiple Access (TDMA). Hence, there are currently three incompatible digital standards operating in the U.S. market, where incompatible means that users of one standard cannot roam to the other. Both TDMA and CDMA have near national coverage, while GSM's coverage is quite extensive also.

Funk (1998) provides some qualitative evidence that firms are likely to dominate domestically, where the "home" standard is dominant, that is firms by and large have not succeeded in markets with "non-domestic" standards. Funk (1998) quotes Yano research estimates that Motorola's share of the world market dropped from 40% to 32% from 1994 to 1995, as the installed base on GSM grew. He cites Nomura Research Institute estimates that Motorola's 1996-1997 average market shares were 23% for GSM subscribers and 70% for AMPS subscribers. Hence Motorola dominated in the U.S., while Nokia/Ericcson dominated in Europe.

It is interesting to note that 63% of all mobile phones sold in the U.S. are produced domestically. In the case of audio equipment, domestic U.S. manufacturers have only 19% of the U.S. market; similarly, domestic U.S. manufacturers of video equipment have only 26% of the U.S. market.[16] While this difference may be due in part to the fact that the consumer electronics industry is more mature than the mobile phone industry, it is possible that incompatibilities in the mobile phone networks delayed (reduced) international trade. This may be difficult to test empirically. In any case, it would be interesting to quantify the trade flows and the entry (or lack of entry) of developing country firms into this industry.

Conclusion

This chapter has tried to lay out a framework for quantifying the trade impact of compatibility standards from an industrial organization perspective. The paper also discussed several industries that provide natural experiments to measure the effect of compatibility standards on international trade. The next step is to begin collecting the necessary data.

NOTES

1. See the references cited within Baseman, Warren-Boulton, and Woroch (1995).

2. Park (1997) cites a 1982 *Consumer Reports* publication that tested various VCR models. The report concluded that there was no significant difference in the characteristics or qualities of the two platforms.

3. See Gabel (1991). The JVC lead was due in part to the fact that JVC cassettes initially had a longer playing time and in the early adoption period (1976-1980), consumers primarily used VCRs to record television programs in order to replay them at a later time.

4. This shift includes developing countries to the extent that they produce or employ these hi-tech products.

5. Unpublished manuscripts that provide empirical evidence of virtual network effects include Park (1997), the VCR market, Shankar and Bayus (1997), the Home Video Game Industry, and Berndt, Pindyck, and Azoulay (1999), the pharmaceutical industry.

6. In Gandal and Shy (2000), governments cannot explicitly choose to restrict international trade. They can, however, reduce the amount of international trade by not recognizing foreign standards. Kubota (2000) examines related issues. In her setting, each country already has an existing incompatible standard. Due to the assumption of homogeneous consumers, if international trade takes place, only one of the international standard survives. The benefit from a single standard is a larger network, while the cost is the "switch" from the domestic to international standard.

7. It's not clear whether the OECD's use of the word "standard" means largely undifferentiated in terms of demand characteristics, or whether "standard" means electronics products which are compatible.

8. This is because under incompatibility, a firm that loses a sale of a particular component loses the sales of all other components of its system. This makes firms price more aggressively.

9. See Crane (1979) & Pelkans and Beuter (1987).

10. See Burton and Saelens (1987).

11. Alternative explanations are that the European television industry began a decade later than the U.S. industry and that European countries adopted policies to keep Japanese television receivers out of their markets.

12. In the case of CD-players, the discs are the complementary products, while in the case of the VCRs, the cassettes are the complementary products.

13. Discussion in this section primarily comes from Gandal, Greenstein, and Salant (1999) and the sources cited within.

14. The other dominant operating system was the proprietary Apple II operating system. Apple served a hobbyist market from the beginning and never successfully moved into the business market, as did CP/M and DOS. See Gabel (1991), Langlois (1992), and Gandal, Greenstein, and Salant (1999).

15. Since Microsoft kept the rights to the MS-DOS/PC-DOS operating system that it developed under contract for IBM, Microsoft was able to sell OEM (Original Equipment Manufacturer) licenses of MS-DOS/PC-DOS to other computer manufacturers.

16. These data come from the Consumer Electronics Association.
See http://www.ce.org/index.

REFERENCES

Barrett, S. 1994. Strategic Environmental Policy and International Trade. *Journal of Public Economics* 54: 25-38.

Baseman, K., F. Warren-Boulton, and G. Woroch. 1995. Microsoft Plays Hardball: The Use of Exclusionary Pricing and Technical Incompatibility to Maintain Monopoly Power in Markets for Operating System Software. *Antitrust Bulletin* XL(2): 265-315.

Berndt, E., R. Pindyck, and P. Azoulay. 1999. Network Effects and Diffusion in Pharmaceutical Markets: Antiulcer Drugs. mimeo. Cambridge MA: Massachusetts Institute of Technology.

Boom, A. 1995. Asymmetric International Minimum Quality Standards and Vertical Differentiation. *Journal of Industrial Economics* 43: 101-19.

Burton, F., and F. Saelens. 1987. Trade Barriers and Japanese Foreign Direct Investment in the Colour Television Industry. *Managerial and Decision Economics* 8: 285-293.

Casella, A. 1996. On Standards and Trade: A Review of Simple Results. In *Fair Trade & Harmonization*, eds. J. Bhagwati and R. Hudec. Cambridge, MA: Massachusetts Institute of Technology Press.

Chou, C. and O. Shy. 1990. Network Effects without Network Externalities. *International Journal of Industrial Organization.* 8: 259-270.

Church, J., and N. Gandal. 1992. Network Effects, Software Provision and Standardization. *Journal of Industrial Economics.* 40: 85-104.

Crane, R. 1979. The Politics of International Standards. Norwood, NJ: Ablex Publishing.

David, P., and S. Greenstein. 1990. The Economics of Compatibility Standards: An Introduction to Recent Research. *Economics-of-Innovation-and-New-Technology.* 1: 3-41.

David, P., and M. Shurmer. 1996. Formal Standards-setting for Global Telecommunications and Information Services. *Telecommunications Policy*, 20: 789-815.

Farrell, J. and G. Saloner. 1985. Standardization, Compatibility and Innovation. *RAND Journal of Economics*. 16: 70-83.

Farrell, J., H. Monroe, and G. Saloner. 1998. The Vertical Organization of Industry: Systems Competition versus Component Competition. *Journal of Economics and Management-Strategy*. 7: 143-82.

Funk, J. 1998. Competition Between Regional Standards and the Success and Failure of Firms in the World-wide Mobile Communications Market. *Telecommunications Policy*. 22: 419-441.

Gabel, L. 1991. Competitive Strategies for Product Standards. New York, NY: McGraw Hill.

Gandal, N. 1994. Hedonic Price Indexes for Spreadsheets and an Empirical Test for Network Externalities. *RAND Journal of Economics*. 25: 160-170.

Gandal, N., S. Greenstein, and D. Salant. 1999. Adoptions and Orphans in the Early Microcomputer Market. *Journal of Industrial Economics*. XL:VII: 87-105.

Gandal, N., M. Kende and R. Rob. 2000. The Dynamics of Technological Adoption in Hardware/Software Systems: The Case of Compact Disc Players. *RAND Journal of Economics*. 31: 43-61.

Gandal, N. and O. Shy. 2001. Standardization Policy and International Trade. *Journal of International Economics.* 53: 363-383.

Gilbert, R. Symposium on Compatibility: Incentives and Market Structure. *Journal of Industrial Economics*. 40: 1-8.

Gordon, R. 1990. The Measurement of Durable Goods Prices. *National Bureau of Economic Research Monograph Series*. Chicago, IL and London: University of Chicago Press.

Greenstein, S. 1992. Did Installed Base Give an Incumbent Any (Measurable) Advantages in Federal Computer Procurement. *RAND Journal of Economics*. 24: 19-39.

Grindley, P. 1995. Standards Strategy and Policy: Cases and Stories. Oxford, U.K.: Oxford University Press.

Grindley, P. and R. McBryde. 1992. Compact Disc and Digital Audio Tape: The Importance of Timing. In *Standards, Business Strategy and Policy, A Casebook*, ed. P. Grindley. Centre for Business Strategy. London: London Business School.

Grindley, P., Salant, and L. Waverman. 1999. Standard Wars: The Use of Standard Setting as a Means of Facilitating Cartels Third Generation Wireless Telecommunications Standard Setting. mimeo.

Jensen, R., and M. Thursby. 1996. Patent Races, Product Standards, and International Competition. *International Economic Review*. 37:21-50.

Katz, M. and C. Shapiro. 1985. Network Externalities, Competition, and Compatibility. *American Economic Review*. 75: 424-440.

———. 1986. Technology Adoption in the Presence of Network Externalities. *Journal of Political Economy*. 94: 822-841.

———. 1994. Systems Competition and Network Effects. *Journal of Economic Perspectives,* 8: 93-115.

Kubota, K. 2000. Trade Negotiations in the Presence of Network Externalities. mimeo.

Langlois, R., and P. Robertson. 1992. Innovation in Modular system: Lessons from the Microcomputer and Stereo Component Industries. *Research Policy*, 21: (4) 297-313.

Kende, M. 1991a. Strategic Standardization in Trade with Network Externalities. INSEAD Working Paper 92/50/EP.

———. 1991b. Gains from Standardization: the Case of Numerical Controls. INSEAD Working Paper 92/51/EP.

Lutz, S., 1996. Does Mutual Recognition of National Minimum Quality Standards Support Regional Convergence. CERP Discussion Paper # 1385.

Matutes, C. and P. Regibeau. 1988. Mix and Match: Product Compatibility Without Network Externalities. *RAND Journal of Economics*, 19: 221-234.

———. 1996. A Selective Review of the Economics of Standardization: Entry Deterrence, Technological Progress, and International Competition. *European Journal of Political Economy*, 12: 183-206.

McGahan, A. 1991. Phillips Compact Disc Introduction (B). Harvard Business School Case Study no. 792-036.

Park, S. 1997. VHS/Beta and Network Externalities. Department of Economics, mimeo. Stonybrook, NY: State University of New York.

Pelkans, J., and R. Beuter. 1987. Standardization and Competitiveness: Private and Public Strategy. In *Product Standardization and Competitive Strategy*, ed. L. Gabel. Amsterdam: Elsevier Science Pubilshers.

Saloner, G. and A. Shepard. 1995. Adoption of Technologies with Network Externalities: An Empirical Examination of the Adoption of Automated Teller Machines. *RAND Journal of Economics*, 26: 479-501.

Shankar, V., and B. Bayus. 1997. Network Effects and Competition: An Empirical Analysis of the Home Video Games Industry. mimeo.

Shy, O. 1991. International Standardization and Protection. Sackler Institute of Economics Studies, Discussion Paper 11-91. Tel Aviv: Tel Aviv University.

Sykes, A. 1995. Product Standards for Internationally Integrated Good Markets. Washington, D.C.: Brookings Institution.

CHAPTER 6

GMOs, Trade Policy and Welfare in Rich and Poor Countries

Chantal Nielsen
Kym Anderson

Introduction

The use of modern biotechnology to create genetically modified organisms (GMOs) through agricultural production is seen as exciting and valuable by some people, while others are objecting strongly to its use. Concerns have been raised about the safety of consuming genetically modified foods, the environmental impact of growing genetically engineered plants, and the ethics related to using the technology *per se*. Skepticism toward genetic engineering has been particularly outspoken in Western Europe, and the development and use of genetically engineered products there has been stunted as a result. In contrast, farmers in North America and several large developing countries such as Argentina and China have adopted genetically modified crops as they have become available, and citizens there have generally accepted that development.

In Western Europe where food supplies are abundant and incomes are high, people can afford to be critical about the introduction of new agricultural technologies and production processes about which they are unsure. In developing economies, by contrast, the benefit/cost ratio is very different. Many food-insecure people in developing countries live in rural areas, earn a significant share of their income from agriculture, and meet a substantial share of their food needs from their own production. For them, increasing agricultural productivity and thereby real income is a high priority. And for the urban poor in those countries, anything that lowers the effective price of basic foods is highly desirable. Given the large value shares of agriculture in production and food in consumption in developing economies, GMO technology has the potential to generate significant economy-wide benefits that may well dwarf any costs that may be perceived in those countries in terms of environmental and food safety risks.

The extent to which developing countries can reap those potential net benefits from the new biotechnology is not independent of consumer and policy responses to GMOs in the developed countries, however. In so far as developing countries are dependent on exports of farm products to richer (potentially GMO-critical) countries, they must consider how GMO policy actions in those countries might affect market access conditions, the agrifood demand structure and hence world market prices. There are basically three possible types of technical barrier to trade: import bans, technical standards and information remedies. Standards concerning GMOs have not yet been established, but there is clearly a demand for better information through labelling, and import bans also are under consideration by some groups. France and other European countries, for example, claim they have the right to adopt the precautionary principle given the scientific uncertainties associated with GM use – an argument that may be strengthened if the Biosafety Protocol negotiated in Montreal in January 2000 is ratified. On the other hand, countries that are generally positive about GM use, such as the United States, might accuse the Europeans of using this issue as an excuse for replacing price-support policies, which are being phased down following the Uruguay Round agreements, with technical barriers to trade (Sykes 1999). Hence trade in agricultural biotech products is likely to become a controversial issue within the World Trade Organization (WTO), with potentially adverse consequences for both developed and developing country agricultural-exporters. Even if these countries are fortunate enough to benefit from the new technology in producing for their own consumption, they might risk having their products refused at the European borders solely because they are genetically modified.

This chapter starts by providing a brief overview of the current status of transgenic crops in agriculture, and an exploration of the potential for GMOs in developing countries. It then elaborates on ways in which the emergence of GMOs is prompting the development of national regulations and international agreements that could raise difficulties and potentially lead to trade disputes in the WTO. A major point of contention could be the requirement that scientific evidence is needed under the SPS Agreement to justify a trade restriction: many in Europe argue that the precautionary principle should operate where there is uncertainty as to the effects on health and/or the environment, even if there is no evidence yet of harm. An empirical model of the global economy (the GTAP model) is then used to analyze the possible effects on world production and trade patterns and on national economic welfare of selected (non-European) countries adopting genetically modified (GM) crops, first without and then with particular European policy and consumer responses. Specifically, the effects of an assumed degree of productivity growth in the maize and soybean sectors in selected countries are explored, and those results are then compared with what they would be if Western Europe chose to ban consumption and hence imports of those products from countries adopting GM technology. The effects of an alternative market-based shift in consumer preferences are then compared with this regulatory approach. The implications for developing countries' participation in world agricultural and food trade are discussed. The final section of the chapter discusses the areas in which future research on the economics of GMOs and related national and international policy responses might focus.

Genetic Engineering in Agriculture

The most recent research and development advances in modern biotechnology have introduced an ever-widening range of genetically engineered products to agriculture. While traditional biotechnology improves the quality and yields of plants and animals through, for example, selective breeding, genetic engineering[1] is a new biotechnology that enables direct manipulation of genetic material (inserting, removing or altering genes). In this way the new technology speeds up the development process, shaving years off R&D programs. Protagonists argue that genetic engineering entails a more-controlled transfer of genes because the transfer is limited to a single, or just a few selected genes, whereas traditional breeding risks transferring unwanted genes together with the desired ones. Against that advantage, antagonists argue that the side effects in terms of potentially adverse impacts on the environment and human health are unknown.

Genetic engineering techniques and their applications have developed rapidly since the introduction of the first genetically modified plants in the 1980s. In 1999 genetically modified crops occupied 40 million hectares of land – making up 3.4% of the world's total agricultural area[2] and representing a considerable expansion from less than 3 million hectares in 1996. Cultivation of transgenic crops has so far been most widespread in the production of soybeans and maize, accounting for 54% and 28% of total transgenic crop production in 1999, respectively (James 1997, 1999). The United States holds almost three-fourths of the total crop area devoted to genetically modified crops. Other major GM-producers are Argentina, Canada and China. At the national level, the largest shares of genetically engineered crops in the total in 1999 were found in Argentina (approximately 90% of the soybean crop), Canada (62% of the rapeseed crop) and the United States (55% of cotton, 50% of soybean and 33% of maize) (James 1999).

Continued expansion in the use of transgenic crops will depend in part on the benefits obtained by farmers cultivating transgenic instead of conventional crops relative to the higher cost for transgenic seeds[3]. To date genetic engineering in agriculture has mainly been used to modify crops so that they have improved agronomic traits such as tolerance of specific chemical herbicides and resistance to pests and diseases. Development of plants with enhanced agronomic traits aims at increasing farmer profitability, typically by reducing input requirements and hence costs. Empirical data on the economic benefits of transgenic crops are still very limited, however. The effects vary from year to year and depend on a range of factors such as crop type, location, magnitude of pest attacks, disease occurrence and weed intensity.

In developing countries one of the main reasons for low crop yields is the prevalence of biotic stresses caused by weeds, pests and diseases. Transgenic crops, which have been modified to exhibit traits such as herbicide tolerance or pesticide resistance, can provide protection against several of these. The development of more complex traits such as drought resistance is underway and highly relevant for tropical crops that are often growing under harsh natural and weather conditions and on poor-quality soils. Furthermore, the development of foods with enhanced nutritional value – such as the recent development of 'Golden Rice' that is enhanced with Vitamin A and iron – may be a low-cost way of dealing with widespread malnutrition problems. There are not many estimates of the potential productivity impact that widespread cultivation of transgenic crops may have in the developing countries but, according to James and Krattiger (1999 p.1), "[a] World Bank panel has estimated that transgenic technology can increase rice production in Asia by 10 to 25 percent in the next decade." Thus modern biotechnology may well be able to ease food supply constraints and food insecurity problems in the developing world significantly, just as the dwarf cereal varieties did in the so-called Green Revolution of the 1960s and 1970s. But the extent to which the potential of this new green revolution is realized will depend in non-trivial ways on GMO preferences and policy developments in rich countries.

National Regulations

Given the novelty of commercial application of genetic engineering techniques, the regulatory frameworks around the world are in the process of being formulated and re-formulated in response to e.g. consumer reactions to these new products. Existing regulations differ substantially both in scope and stage of implementation, varying from very restrictive regulations in certain industrial countries to non-existent in many developing countries. The resistance of consumer and environmental groups in the European Union (EU) to genetically modified foods and to the use of GMOs in agricultural production has triggered the imposition of a *de facto* moratorium on the authorization of new releases of GMOs as of June 1999. This could be a prelude to a future EU ban on the cultivation of genetically modified crops and on imports of foods containing GMOs.

Before the imposition of the current moratorium, *all* deliberate releases of GMOs in the European Union from laboratory testing, to field testing and final marketing. By contrast, the permit procedure in the United States is more flexible. Permits for field-testing and release into the environment of transgenic crops must be obtained from the Animal and Plant Health Inspection Service (APHIS) of the US Department of Agriculture. But for transgenic crops with which APHIS has experience, a notification system enables a more rapid permitting procedure (Nelson et al. 1999).

With regard to labeling, the United States Food and Drug Administration (FDA) does not require labeling of genetically modified foods solely because they are genetically modified. Only if the transgenic food differs substantially from its conventional counterpart will a pre-market permit and labeling be required. This is because the FDA does not distinguish between foods produced from genetically modified crops and foods produced from crops developed by other technologies, and so genetically engineered foods and food ingredients are 'only' required to meet the same safety standards as other food products (Food and Drug Administration 1995).

The EU, by contrast, requires labeling of all foodstuffs, additives and flavors containing 1% or more genetically engineered material (Regulations 1139/98 and 49/2000). This *de minimis* level takes into account the fact that a certain degree of co-mingling is unavoidable with common handling and transportation systems in the agri-food process and therefore does not require labeling of food ingredients that unintentionally contain GMOs if this level does not exceed 1%. Labeling in the EU is based on scientific evidence proving the presence of genetically engineered DNA or protein. There are loopholes in these labeling requirements, however. Processed foods whose ingredients have been genetically modified, but where the food production process has eliminated the external DNA or protein, do not need to be labeled because the genetically engineered material cannot be identified in the final product. In response, retailers in EU countries, such as Denmark, are taking their labeling requirements one step further by requiring suppliers to label their products not only if GMO presence can be verified scientifically, but also if there is a possibility that the product could contain GMOs.

Regulatory changes are also taking place in other countries. South Korea, Japan, Australia, Mexico and New Zealand have recently decided to enact labeling requirement laws for transgenic foods. According to James (1998), many developing countries also are beginning to establish regulations related to genetically engineered products. Operational field-testing regulations have been implemented in, for example, Argentina, Brazil, Mexico, Chile, Costa Rica, Cuba, India, the Philippines and Thailand. Sri Lanka has taken a more drastic step by simply banning the imports of GMOs pending further clarification as to their environmental and food safety impacts.

Demand for labeling of food products is of course not a new phenomenon. As consumers become richer they tend to demand to be able to make qualified consumption choices based on high levels of information and reliable labeling systems. A management system that preserves the identity of a crop in terms of its origin or other characteristics, and hence enables reliable labeling of food products, is known identity preservation (IP). Such systems are used today to classify almost all traded agricultural and food products according to certain criteria – some more sophisticated than others. These systems incur real resource costs because they require specific actions and increased administration pre-farm, on the farm, and "at all stages in transport, storage, processing, food manufacture, distribution and retailing" (Buckwell et al. 1999 p.vii). Tests and tolerance levels are important factors determining the size of these costs. Tests are often used to verify a specific physical or chemical content. In cases where such tests are not possible, e.g. when the criteria is related to the production process rather than the product characteristics, credibility of the IP system relies on the integrity of the supplier and government control. When guarantees cannot be made about the 'cleanliness' of a product (e.g. 100% GMO-free) the question of tolerance levels arises. Generally, the lower the tolerance level, the higher the cost. How the additional costs relating to IP and labeling are distributed between suppliers, farmers, food processors, distributors, and final consumers depends on the size of the price elasticity of demand at each stage. The lower the price elasticity of demand (e.g. due to the absence of close substitutes), the larger share of the costs the consumers will bear, ceteris paribus.

Although labeling entails higher costs for producers, they would willingly bear them if a sufficiently higher product price resulted. Quality-enhanced products – more nutritious foods, crops with better processing characteristics, organic foods, and perhaps also foods guaranteed free of GMOs – may attract such a premium price. For genetically modified crops that 'only' exhibit agronomically improved characteristics, however, the situation is different. Most users do not view such modified products as higher quality where the specific aim of this type of technology is to reduce farmers' costs and associated damage to the environment from farm chemicals. In the absence of an incentive for producers to voluntarily identify their product with a label saying it "may contain GMOs", positive labeling regulations would have to be established if this is what consumers and policy makers demand. The costs of such regulation would be non-trivial, not least because it would require identity preservation throughout the food chain for what would amount to most processed foods. The benefits, on the other hand, are far from obvious, since the label has almost no information content because there are no known risks from consuming GMOs (in contrast to, for example, the positive health warning label on cigarette packets).

A non-regulatory alternative to positive labeling regulations is simply to allow the use of negative labels such as "this product contains no GMOs"[4]. That begs the question of defining the threshold below which 'no' applies, and presumably agreeing on that multilaterally for it to be meaningful for internationally traded products. And it still requires non-GM products to be kept separate from GM-inclusive products. However, since the vast majority of processed foods contain some GMOs now, this alternative requires labels on a much smaller proportion of products. Also, like organic food, those products reputedly are the ones capable of attracting a higher price to pay for their identity preservation.[5]

International Trade Agreements

The Cartagena Protocol on Biosafety

Given the different national approaches to regulation of genetically modified products, future trade disputes are a distinct possibility (Kelch et al. 1998). With the objective of ensuring safe transboundary movement of living modified organisms resulting from modern biotechnology, the Cartagena Protocol on Biosafety was finalized in Montreal on 29 January 2000. The Biosafety Protocol, if ratified by the parliaments of 50 signatories, will allow governments to decide whether or not to accept GMO imports and under what conditions, and reconfirms their rights to set their own domestic regulations. Most importantly, the Protocol stipulates that lack of scientific evidence regarding potential adverse effects of GMOs on biodiversity, taking also into account risks to human health, shall not prevent a signatory from taking action to restrict the import of such organisms in order to avoid or minimize risks (UNEP 2000). In essence, this reflects an acceptance of the guiding influence of the precautionary principle[6], that is, "better safe than sorry". In terms of documentation, the Protocol requires that GMOs intended for intentional introduction into the environment or for contained use clearly identifies them as living modified organisms. But for modified organisms intended for direct use as food or feed, or for further processing, all that is required is a label stating that the product "may contain" such organisms. No labeling requirements for processed foods such as cooking oil or meal were established. Hence the Protocol does not address growing demands by hard critics of biotech who call for labeling of products if genetic engineering techniques have been used at any stage in their production process regardless of whether or not this can be verified in the final product through testing.

The Cartagena Protocol on Biosafety is a part of the 1992 Convention on Biological Diversity to which in particular the United States is not a signatory. Nevertheless, it is expected that the US will abide to the rules of this Protocol so as to facilitate trade. However, there is an important aspect of the Protocol, which seems open to interpretation and hence potential dispute. This is the relationship between the Protocol and the WTO agreements. The text (UNEP 2000 p.1) states that the "Protocol shall not be interpreted as implying a change in the rights and obligations of a Party under any existing international agreements" while at the same time, the latter statement is "not intended to subordinate [the] Protocol to other international agreements." Given the existing dispute settlement mechanism of the WTO, it is likely that potential disputes will be taken up in this forum. The Protocol has the objective of protecting and ensuring sustainable use of biological diversity while also taking into account risks to human health. Agreements within the WTO (e.g., GATT Article XX, SPS Article 2) also acknowledge the rights of a country to protect its environment and ensure food safety for consumers. But WTO members are also obliged to adhere to agreements that restrict the way trade-restrictive measures are used to achieve these goals. Hence some of the current WTO agreements may prove to be in conflict with the rights to restrict trade in living modified organisms provided for in the Protocol.

The WTO Agreements and Transgenic Products

The key goal of the WTO is to achieve effective use of the world's resources by reducing barriers to international trade. The WTO does, however, acknowledge the need of member states to apply and enforce trade-restricting measures in order to protect human, animal or plant health and life as well as public morals. That right for a country to set its own environmental and food safety regulations at the national level is provided for in Article XX of the GATT. Use of such measures must be consistent with the key principles of the WTO, however: non-discrimination among member states, 'national treatment' of imports once having entered the domestic market, and transparency of customs procedures[7].

Members of the WTO also have trade obligations under other WTO agreements that restrict the extent to which trade measures can be used against GMOs without risking a case coming before the WTO's Dispute Settlement Body. More specifically related to food safety and animal and plant health are the Agreement on Sanitary and Phytosanitary measures (SPS) and the Agreement on Technical Barriers to Trade (TBT). These agreements allow member states to impose certain restrictions on trade if the purpose of the measure is to protect human, animal or plant life and health. The TBT agreement also covers technical measures aimed at protecting the environment and other objectives. At the same time the agreements aim to ensure that applied measures and technical regulations do not pose unnecessary or arbitrary barriers to trade and that they are no more trade-restrictive than necessary to fulfill the legitimate objective (WTO 1995 and 1998a, b).

Both the SPS and TBT agreements encourage the use of international standards, guidelines and recommendations where they exist, such as the Codex Alimentarius. Currently there are no international standards for genetically modified products although the Biosafety Protocol explicitly notes that signatories "shall consider the need for and modalities of developing standards with regard to the identification, handling, packaging and transport practices, in consultation with other relevant international bodies." (UNEP 2000 p.10, Article 18.3). International harmonization of regulatory approval procedures for genetically modified products is being discussed in several forums including the FAO and OECD (1998). The establishment of international standards for the production, regulation and labeling of these products may be helpful as a way of reducing future trade disputes. Nevertheless, under the SPS agreement a country may apply higher standards *only* if these can be justified by appropriate scientific risk assessments. In other words, the SPS agreement explicitly allows member states to set their own standards for food safety and animal and plant health, but at the same time requires that measures be based on scientific risk assessments. The TBT agreement is more flexible because member states can decide that international standards are inappropriate for a number of other reasons, such as national security interests (GATT Article XXI). Hence determining which agreement a given trade measure is covered by is of key importance. The SPS agreement covers food safety measures and animal and plant health standards regardless of whether or not these are technical requirements. The TBT agreement, on the other hand, covers all technical regulations, voluntary standards and compliance procedures, except when these are sanitary and phytosanitary measures as defined in the SPS agreement (WTO 1998a).

The scientific requirement of the SPS agreement is important because it provides for a more objective approach than the TBT agreement in determining what is a justified trade restriction and what is hidden protectionism. On the other hand, the SPS agreement may be inadequate for tackling restrictions introduced on the basis of e.g. voter sentiment in relation to food production methods such as genetic engineering. Official disputes about trade in genetically modified products have not yet materialized, but experience from earlier WTO dispute settlement cases that are comparable to the GMO debate give an indication as to how the existing rules may be applied. The SPS agreement was used in the beef hormone dispute between the US and the EU, for example (WTO 1998c). In short, the EU import ban on meat and meat products from hormone-fed livestock was found to be in conflict with the EU's WTO obligations, the main argument being that the EU could not present documented scientific risk assessment to justify the ban.

Scientific evidence is not always sufficient for governments to make policy decisions, or it may simply be unavailable. In such cases, Article 5.7 of the SPS agreement allows WTO member states to take precautionary measures based on available pertinent information. At the same time, members are obliged to seek additional information so that a more objective evaluation of the risks related to the relevant product or process can be made within a reasonable period of time. The precautionary principle is an understandable approach to uncertainties about genetically modified products, but there is a risk that when used in connection with internationally traded products, it can be captured by import-competing groups seeking protection against any new technology-driven competition from abroad. It may thus be extremely difficult to assess whether a measure is there for precautionary reasons or simply as a form of hidden protectionism. For this reason, how the provisions of the Biosafety Protocol – the most explicit acceptance of the use of the precautionary principle in an international trade agreement relating to agri-food products to date – will be interpreted in light of the existing WTO commitments is a moot point.

The existing trade agreements deal with regulations and standards concerning products as well as production processes and methods *if they affect the characteristics or safety of the product itself*. Standards for production processes that do *not* affect the final product are not covered by the existing agreements. In relation to genetically engineered products, if the process itself alters the final product in a way so that there are adverse environmental or health effects associated with consumption, use or disposal of the product, restricting trade in this product would not violate existing WTO rules, *ceteris paribus*. If genetic engineering only concerns the production process and not the final characteristics of a transgenic product, domestic regulations that restrict the use of this method of production cannot be used to restrict imports of products produced by this method simply because the importing country finds it unacceptable by its own environmental, ethical or other norms.[8]

This discussion leads back to the role of scientific evidence. Some would argue that genetically modified products are different from conventional products *regardless* of whether or not this can be verified scientifically in the final product. One of the priorities of the European Commission in the next WTO round of multilateral trade negotiations is to obtain a clarification of the role of non-product-related processes and production methods within the WTO (European Commission 1999). If trade restrictions based on production methods are allowed, this could lead to the inclusion of a long list of non-tariff barriers not only in relation to biotechnology products, but also related to labor standards, for example. Hence this move is being strongly resisted by developing countries.

Labeling of foods in relation to international trade is normally covered by the TBT agreement unless the label relates directly to food safety, in which case it is covered by the SPS agreement. The WTO's labeling rules encourage the use of international standards but, as discussed earlier, they do not yet exist for transgenic foods. Related to the discussion above about 'products' and 'processes', only labeling programs that concern production processes affecting the final product would be covered by the existing TBT agreement. Determining whether or not a genetic modification affects the final product will probably have to be done on a case-by-case basis, but to the extent that a labeling program is covered by the TBT agreement, the WTO requires that it does not pose an unnecessary barrier to international trade.

Where labeling programs are not encompassed by the TBT agreement, which potentially may be the case for many transgenic products, the other agreements of the WTO will be applicable without exceptions (Tietje 1997). GATT Article III concerning non-discrimination, for example, stipulates that member states may not discriminate between otherwise like goods on the basis of their country of origin. A key issue using this Article will be the interpretation of the concept of 'like goods' and whether the presence of genetically modified material is 'sufficient' to differentiate products. Article III seeks to avoid measures that are based on a false differentiation of products, as determined on a case-by-case basis.

In response to the heated debate about transgenic crops and foods, markets for guaranteed non-GMO maize and soybeans have already begun to develop. Several large food companies around the world have a non-GMO policy in place. Clearly, consumer demand and willingness to pay will determine the viability and size of this market, but producers of GMO-free products would be interested in labeling their products as such in order to distinguish them from GMO-inclusive products. Even if such labeling is based on voluntary decisions made by private companies, public authorities will often control the use of the label. Possible trade-distorting effects can thus be argued to be the responsibility of the state (i.e. the WTO member) and thus subject to WTO rules. The key requirement on the part of the WTO is that labeling does not discriminate between domestic and foreign products. Summarizing, the emergence of GMOs in agricultural and food production introduces several issues to be dealt with by the WTO membership:

1. The aim of national regulation: whether it is for protecting the environment and/or human health will influence which WTO agreements a given regulation will fall under. The distinction may prove to be important in terms of interpreting the Biosafety Protocol and its relation to the WTO rules, since TBT rules are less strict than rules under the SPS agreement.
2. The SPS requirement of scientific evidence: it now stands in sharp contrast to the Biosafety Protocol where it is explicitly stated that lack of such evidence need not prevent importing countries from taking action.
3. The product/process distinction: the Biosafety Protocol implicitly expresses a distinction between products based on their production process, i.e. with or without genetic engineering, unlike WTO rules on 'like products'.
4. Mandatory labeling requirements of all GMO-inclusive products including processed foods: as long as they do not discriminate between foreign and domestic goods they will probably not violate existing WTO rules, even if they add significantly to trade costs.

Estimating Economic Effects of GMO Adoption and of Policy or Consumer Reactions

The above institutional considerations and apparent differences in preferences and views on consumers' right to know are unlikely to disappear in the foreseeable future. The extent to which that could lead to trade disputes depends heavily on the directions and magnitudes of the production, trade and welfare consequences of different responses for different countries. Theory alone is incapable of determining even the likely direction, let alone the magnitude, of some of the effects of various policy or consumer responses to GMO adoption, so an empirical modeling approach is called for, along with assumptions about the size of various productivity and preference changes. What follows is an early attempt at doing that for maize and soybeans.

Global Modeling Framework

This section examines empirically the production, trade and welfare effects of GM crop adoption by selected regions, first without and then with specific policy and consumer responses in Western Europe. This is done using an applied analytical framework involving a global economy-wide model and database known as GTAP (Global Trade Analysis Project).[9] Being a general equilibrium model, GTAP describes both the vertical and horizontal linkages between all product markets both within the model's individual countries and regions as well as between countries and regions via their bilateral trade flows. The database used for this application reflects the global economic structures and trade flows of 1995 (GTAP database Version 4 - see McDougall et al. 1998) and has been aggregated to 16 regions and 17 sectors, as shown in Table 6.1. This aggregation highlights the main participants in the GMO debate, including key groups of developing countries, and the main primary agricultural sectors affected and their related processing industries. That includes not only maize and soybeans for the present analysis, but also rice and cotton, which are analyzed in Nielsen and Anderson (2000b).

Scenarios[10]

Currently it is maize and soybeans that benefit most from GM-technology. The scenarios analyzed here assume that GM-driven productivity growth occurs only in the following GTAP sectors: cereal grains (excluding wheat and rice) and oilseeds. Detailed empirical information about the impact of GMO technology in terms of reduced chemical use, higher yields and other agronomic improvements is at this stage quite limited (see e.g. OECD (1999) and Nelson et al. (1999)). Available empirical evidence (e.g. USDA 1999 and James 1997, 1998) suggests that cultivating GM crops has general cost-reducing effects. The scenarios analyzed here are therefore based on a simplifying assumption that the effect of adopting GM crops can be captured by

TABLE 6.1. Regions and Sectors in the GTAP Model Analysis

1. Australia & New Zealand	11. Rest of Latin America	1. Paddy rice	11. Meat and dairy products
2. Japan	12. Western Europe	2. Wheat	12. Vegetable oils and fats
3. Other high-income Asia	13. Central Europe & FSU	3. Cereal grains	13. Processed rice
4. China	14. South Africa	4. Vegetables, fruits & nuts	14. Other Processed food
5. Rest of East Asia	15. Sub-Saharan Africa	5. Oilseeds	15. Textiles and clothing
6. India	16. Middle East & North Africa	6. Plant-based fibers	16. Other manufactures
7. Rest of South Asia		7. Other crops	17. Services
8. North America		8. Livestock	
9. Mexico		9. Forestry & fishing	
10. Southern Cone		10. Energy & minerals	

Source: Author's listing.

a Hicks-neutral technology shift, i.e. a uniform reduction in all inputs to obtain the same level of production. For present purposes the GM-adopting sectors are assumed to experience a one-off increase in total factor productivity of 5%, thus lowering the supply price of the GM crop[11]. Assuming sufficiently elastic demand conditions, the cost-reducing technology will lead to increased production and higher returns to the factors of production employed in the GM-adopting sector. Labor, capital and land will consequently be drawn into the affected sector. As suppliers of inputs and buyers of agricultural products, other sectors will also be affected by the use of genetic engineering in GM-potential sectors through vertical linkages. Input suppliers will initially experience lower demand because the production process in the GM sector has become more efficient. To the extent that the production of GM crops increases, the demand for inputs by producers of those crops may actually rise despite the input-reducing technology. Purchasers of primary agricultural products such as grains for livestock feed will benefit from lower prices, which will affect market competitiveness of livestock products.

The widespread adoption of GM varieties in certain regions will affect international trade flows depending on how the crop in question is traded, and whether or not this trade is left to take place without restrictions related to their GM-characteristic. To the extent that trade is not further restricted and not currently subject to binding tariff quotas, world market prices for these products will have a tendency to decline and thus benefit regions that are net importers of these products. For exporters, the lower price may or may not boost their trade volume, depending on price elasticities in foreign markets. Welfare in the exporting countries would go down for non-adopters but could also go down for some adopters if the adverse terms of trade change is sufficiently strong, hence the need for empirical analysis.

Three scenarios are considered below: The first scenario is a base case with no policy or consumer reactions to GMOs, while the second and third scenarios impose on this base case a specific policy or consumer response. The base scenario examines the implications of widespread adoption of GM maize and soybeans in a number of current and potential biotech front-runner countries/ regions: North America, Mexico, the Southern Cone region of Latin America, India, China, East Asia's other lower-income countries, and South Africa. The countries of Western Europe and elsewhere are assumed to refrain completely from the use of GM crops in their production systems. For the EU this may be interpreted as an extension of the *de facto* moratorium that has been in place there since June 1999. Most notably among the developing countries, Sub-Saharan Africa is assumed to be unable to take advantage of the new technology. As mentioned above, the technology change is assumed to boost productivity growth by 5% in the adopting sectors. Moreover, consumers are assumed not to be concerned about the introduction of GM crops in the agri-food system, so genetically modified and conventional crops are produced side-by-side, and traded in one commingled market. In this base scenario there are no restrictions on trade in genetically modified products.

In the second scenario, Western Europe not only refrains from using GM crops in its own domestic production systems, but the region also is assumed to reject imports of genetically modified oilseeds and cereal grains from GM-adopting regions. It is assumed that the labeling requirements of the Biosafety Protocol enable Western European importers to identify such shipments and that basically all oilseed and cereal grain exports from GM-adopting regions will be labeled "may contain GMOs". So the distinction between GM and GM-free products is simplified to one that relates directly to the country of origin[12]. Furthermore, given the formulation of the labeling requirement in the Biosafety Protocol, it may not be very costly for producers to place such a label on products sold to sensitive markets and hence such labeling costs are ignored. This import ban scenario reflects the most extreme application of the precautionary principle within the framework of the Biosafety Protocol. The welfare consequences of this policy response could go either way for both importing and exporting countries, depending on the size of the terms of trade effects and the bilateral trade structures. This again underscores the need for empirical analysis.

In the event that this is found to be in conflict with the WTO rules, a Western European import ban may be ruled out. The final scenario therefore considers the case in which consumers express their preferences through market mechanisms rather than through government regulation. The scenario analyses the impact of a partial shift in Western European preferences away from imported cereal grains and oilseeds and in favor of domestically produced crops.[13] This is implemented as an exogenous 25% reduction in final consumer and intermediate demand[14] for all imported oilseeds and cereal grains, i.e., not only those which can be identified as coming from GM adopting regions. This can be interpreted as an illustration of the incomplete information provided about imported products (still assuming that GM crops are not cultivated in Western Europe), if a label only states that the product "may contain GMOs". Such a label does not resolve the information problem facing the most critical Western European consumers who want to be able to distinguish between GMO-inclusive and GMO-free products. Thus some European consumers and firms are assumed to choose to completely avoid products that are produced outside of Western Europe. Their demand is thus shifted in favor of domestically produced goods. Western European producers and suppliers are assumed to be able to signal – at no (significant) additional cost – that their products are GM-free by e.g. labeling their products by country of origin. This is possible because it is assumed that no producers in Western Europe adopt GM crops (perhaps due to government regulation), and hence such a label would be perceived as a sufficient guarantee of the absence of GMOs.

Empirical results

Scenario 1: Selected Non-European Regions Adopt GM Cereal Grains and Oilseeds

Tables 6.2 (a) and (b) report the results of the first experiment for selected regions, with Western Europe and Sub-Saharan Africa as non-adopters and the other reported regions (North America, Southern Cone, China, India and South Africa) as adopters of GM cereal grains and oilseeds. A 5% reduction in overall production costs in these sectors leads to increases in cereal grains production of between 0.4% and 3.8%, and increases in oilseed production of between 1.1% and 4.6% in the GM-adopting regions. The production responses are generally larger for oilseeds as compared with cereal grains. This is explained by the fact that a larger share of oilseed production as compared with cereal grains production is destined for export markets in all the reported regions except for South Africa, and hence production is not limited to the same extent by domestic demand, which is less price-elastic[15]. Increased oilseed production leads to lower market prices, cheaper costs, and expanded output in the production of vegetable oils and fats. In North America, cereal grains are also used as livestock feed, so the lower feed prices lead to an expansion of the livestock and meat processing sectors.

Due to the very large world market shares of oilseeds from North and South America and cereal grains from North America, the increased supply from these regions causes world prices for cereal grains and oilseeds to decline by 4.0% and 4.5%, respectively. As a consequence of the more intense competition from abroad, production of cereal grains and oilseeds declines in the non-adopting regions. This is particularly so in Western Europe, a major net importer of oilseeds, of which 47% initially comes from North America. Cereal grain imports into Western Europe increase only slightly (0.1%), but the increased competition and lower price are enough to entail a 4.5% decline in Western European production. In the developing countries too, production of cereal grains and oilseeds is reduced slightly.

The changes in India are relatively small compared with e.g. China and the Southern Cone region. This is explained by the domestic market orientation of these sales. As a consequence, the relatively small production increase causes rather substantial declines in market prices for these products, which in turn benefits the other agricultural sectors through vertical linkages. For example, 67% of intermediate demand for cereal grains and 37% of intermediate demand for oilseeds in India stems from the livestock sector.

Global welfare (as traditionally measured in terms of equivalent variations of income) is boosted in this first scenario by US$9.9 billion per year, two-thirds of which is enjoyed by the adopting regions (Table 2(b)). It is noteworthy that all regions (both adopting and non-adopting) gain in terms of economic welfare except Sub-Saharan Africa. Most of this gain stems directly from the technology boost. The net-exporting GM-adopters experience worsened terms of trade due to increased competition on world markets, but this adverse welfare effect is outweighed by the positive effect of the technological boost. Western Europe gains from the productivity increase in the other regions only in part because of cheaper imports; mostly it gains because of the shift of domestic resources out of relatively protected and supported sectors of EU agriculture triggered by keener competition from abroad. The group of other high-income countries, among which are the Asian nations that are relatively large net importers of the GM-potential crops, benefits equally from lower import prices and a more efficient use of resources in domestic production.

Scenario 2: Western Europe Bans the Imports of GM Cereal Grains and Oilseeds

A Western European ban on the imports of genetically modified cereal grains and oilseeds changes the situation rather dramatically, especially for the oilseed sector in North America, which is initially highly dependent on this particular export market. The result of the European ban is not only a decline in total North American oilseed exports by almost 30%, but also a production decline of 10%, pulling resources such as land out of this sector (Table 6.3 (a)). For cereal grains, by contrast, only 18% of North American production is exported and just 8% of those exports are destined for Western Europe. Therefore the ban does not affect North American production and exports of cereal grains to the same extent as for oilseeds, although the production-enhancing effect of the technological boost is nonetheless dampened significantly by the ban's downward pressure on the international price of maize. Similar effects are evident in the other GM adopting regions, except for India, because its production of these particular crops is mostly sold domestically, so India is largely unaffected by market developments abroad.

For Sub-Saharan Africa, which by assumption is unable to adopt the new GM technology, access to the Western European markets when other competitors are excluded expands. Oilseed exports from this region rise dramatically, by enough to increase domestic production by 4%. Western Europe increases its own production of oilseeds, however, so the aggregate increase in oilseed imports amounts to less than 1%. Its production of cereal grains also increases, but not as dramatically because of an initial high degree of self-sufficiency. The shift from imported oilseeds and cereal grains to domestically produced products has implications further downstream. Given an imperfect degree of substitution in production between domestic and imported intermediate inputs, the higher prices on domestically produced grains and oilseeds mean that livestock feed is slightly more expensive (half of intermediate demand for cereal grains in Western Europe stems from the livestock sector). Inputs to the other food processing industries, particularly the vegetable oils and fats sector, also are more expensive. As a consequence, production in these downstream sectors decline and competing imports increase.

Aggregate welfare implications of this scenario as compared with Scenario 1 are substantially different. Western Europe now experiences a decline in aggregate welfare of US$4.3 billion per year instead of a boost of $2 billion (compare Tables 3(b) and 2(b)). Taking a closer look at the decomposition of the welfare changes reveals that adverse allocative efficiency effects explain the decline. Most significantly, resources are forced into producing oilseeds, of which a substantial amount was previously imported. Consumer welfare in Western Europe is reduced in this scenario because, given that those consumers are assumed to be indifferent between GM and GM-free crops, the import ban restricts them from benefiting from lower international prices.[16] All other regions show a gain in welfare and this is particularly so for India, which in addition to being a GM adopter is also screened from the price developments on international markets for these particular GM crops. As non-adopters, the countries in Sub-Saharan Africa are hardly affected, gaining only very slightly from better terms of trade. In particular a higher price is obtained for their oilseed exports to Western European markets.

Two-thirds of the global gains from the new GM technology as measured in Scenario 1 are eroded by the import ban imposed by Western Europe: they fall from $9.9 billion per year to just $3.4 billion (assuming as before that consumers are indifferent between GM-free and GM-inclusive foods). Almost the entire erosion in economic welfare is borne in Western Europe, although the net-exporting adopters (mainly North America and the Southern Cone region) also gain less in overall economic welfare in this scenario as compared with the previous one. Since the non-adopting regions generally purchase most of their imported cereal grains and oilseeds from the North American region, they benefit even more from lower import prices: their welfare is estimated to be greater by almost one-fifth in the case of a Western European import ban.

TABLE 6.2(a). Scenario 1: Selected Non-European Regions Adopt GM Cereal Grains and Oilseeds Production, Price and Trade Effects, % Changes

	North America	Southern Cone	China	India	South Africa	Western Europe	Sub-Saharan Africa
Production							
Cereal grains	2.1	1.6	1.0	0.4	3.8	-4.5	-2.3
Oilseeds	3.6	4.6	1.8	1.1	3.0	-11.2	-1.3
Livestock	0.8	-0.0	0.1	0.4	-0.1	-0.2	-0.1
Meat & Dairy	0.5	0.0	0.1	1.3	0.1	-0.1	-0.1
Veg.oils,fats	1.1	4.5	1.4	0.0	-2.3	-0.9	-1.2
Other foods	0.2	0.1	0.4	1.5	-0.0	-0.1	0.0
Market prices							
Cereal grains	-5.5	-5.5	-5.6	-6.7	-4.9	-0.5	-0.4
Oilseeds	-5.5	-5.3	-5.6	-6.5	-5.0	-1.2	-0.3
Livestock	-1.8	-0.3	-0.4	-1.4	-0.2	-0.3	-0.3
Meat & Dairy	-1.0	-0.2	-0.3	-1.0	-0.2	-0.2	-0.2
Veg.oils,fats	-2.4	-3.1	-2.6	-1.0	-0.5	-0.5	-0.2
Other foods	-0.3	-0.2	-0.5	-1.0	-0.1	-0.1	-0.2
*Exports**							
Cereal grains	8.5	13.3	16.8	37.3	7.3	-11.5	-20.0
Oilseeds	8.5	10.5	8.2	21.5	7.9	-20.5	-26.5
Livestock	8.9	-2.0	-3.3	9.4	-2.3	-1.1	-1.5
Meat & Dairy	4.8	-0.9	-0.9	5.8	0.3	-0.5	-0.2
Veg.oils,fats	5.8	14.3	5.6	-3.8	-7.6	-4.9	-5.3
Other foods	0.2	0.1	1.6	7.6	-0.5	-0.6	0.1
*Imports**							
Cereal grains	-1.6	-4.6	-4.2	-20.5	-1.9	0.1	11.3
Oilseeds	-2.6	-9.2	-1.6	-8.6	-0.9	2.5	16.5
Livestock	-2.1	1.3	-0.9	-5.2	1.9	0.2	0.5
Meat & Dairy	-1.9	0.2	-0.8	-1.7	0.3	-0.0	0.1
Veg.oils,fats	-3.7	-3.6	-1.7	3.1	2.9	1.3	3.4
Other foods	0	-0.1	-0.6	-3.1	0.2	0.1	-0.1

*Includes intra-regional trade.

Source: Author's GTAP model results.

Scenario 3: Partial shift in Western European preferences against GMOs

In this final scenario, it is assumed that there is no import ban but that some Western European consumers have a strong bias against genetically modified crops. Given the paucity of information embodied in the "may contain GMOs" label, recall that preferences in this scenario are assumed to shift away from imported cereal grains and oilseeds in general, not only from declared GMO-producing regions. As the results in Table 6.4(a) reveal, having consumers express their preferences through market mechanisms rather than through a government-implemented import ban has a much less damaging effect on production in the GM-adopting countries. In particular, instead of declines in oilseed production as in Scenario 2 there are slight increases, and production responses in cereal grains are slightly larger. Once again the changes are less marked for India and in part also for China, which are less affected by international market changes for these products. As expected, domestic oilseed production in Western Europe must increase somewhat to accommodate the shift in preferences away from imports, but not to the same extent as in the previous scenario. Furthermore, there are minor price reductions for agri-food products in Western Europe in part because (by assumption) the shift in preferences is only partial, and so GM-indifferent consumers and firms benefit from lower import prices.

TABLE 6.2(b). Selected Economic Welfare Results

	Equivalent Variation (EV)	Decomposition of welfare results, contribution of (US million):		
	US$ million pa	Allocative Efficiency Effects	Terms of Trade Effects	Technical Change
North America	2,624	-137	-1,008	3,746
Southern Cone	826	120	-223	923
China	839	113	66	672
India	1,265	182	-9	1,094
South Africa	35	4	3	27
Western Europe	2,010	1,755	253	0
Sub-Saharan Africa	-9	-2	-9	0
Other high-income	1,186	554	641	0
Other developing and transition econs	1,085	167	286	646
WORLD	9,859	2,756	0	7,108

Source: Author's GTAP model results.

TABLE 6.3(a). Scenario 2: Western Europe Bans Imports of GM Products Production, Price and Trade Effects, % Changes

	North America	Southern Cone	China	India	South Africa	Western Europe	Sub-Saharan Africa
Production							
Cereal grains	0.9	0.0	0.8	0.4	2.4	5.3	-2.2
Oilseeds	-10.2	-3.6	-0.8	0.8	-11.4	66.4	4.4
Livestock	1.2	0.3	0.2	0.4	0.2	-0.8	0.0
Meat & Dairy	0.8	0.3	0.2	1.4	0.2	-0.5	-0.0
Veg.oils,fats	2.4	8.1	1.6	0.1	-2.4	-3.4	0.0
Other foods	0.3	0.4	0.5	1.6	0.1	-0.5	-0.1
Market prices							
Cereal grains	-6.2	-6.0	-5.6	-6.7	-4.9	0.8	-0.0
Oilseeds	-7.4	-6.8	-6.0	-6.5	-5.9	-5.8	0.4
Livestock	-2.2	-0.7	-0.4	-1.4	-0.2	0.5	0.1
Meat & Dairy	-1.3	-0.4	-0.3	-1.0	-0.2	0.3	0.1
Veg.oils,fats	-3.3	-4.0	-2.7	-1.0	-0.6	2.0	0.0
Other foods	-0.4	-0.3	-0.5	-1.0	-0.1	-0.1	0.0
Exports[*]							
Cereal grains	0.3	-2.9	5.0	23.4	4.2	15.9	-13.1
Oilseeds	-28.8	-69.2	-18.4	-8.7	-41.2	167.2	105.0
Livestock	13.7	4.0	-1.4	12.6	2.2	-3.8	-1.8
Meat & Dairy	7.5	2.1	0.1	7.1	2.6	-1.4	0.3
Veg.oils,fats	14.4	26.2	7.0	1.3	-4.1	-15.0	5.8
Other foods	1.5	1.9	2.0	8.0	0.4	-1.4	-0.6
Imports[*]							
Cereal grains	-1.9	-5.3	-2.8	-20.0	-1.2	3.3	13.4
Oilseeds	-5.6	-21.9	3.0	-3.7	-0.7	0.6	22.5
Livestock	-3.2	0.1	0.1	-5.9	0.9	0.9	0.5
Meat & Dairy	-2.8	-0.5	0.8	-1.8	-0.1	-0.2	-0.0
Veg.oils,fats	-7.7	-5.5	-1.7	4.0	3.6	5.5	2.4
Other foods	-0.6	-0.6	-0.8	-2.8	0.0	0.1	0.2

[*] Includes intra-regional trade.

Source: Author's GTAP model results.

In other words, in contrast to the previous scenario, a certain link between EU prices and world prices is retained here because we are dealing with only a partial reduction in import demand. The output growth in Sub-Saharan Africa in Scenario 2, by taking the opportunity of serving European consumers and firms while other suppliers were excluded, is replaced in this scenario by declines: Sub-Saharan Africa loses export shares to the GM-adopting regions.

The numerical welfare results in this scenario are comparable with those of Scenario 1 (the scenario without the import ban or the partial preference shift) for all regions except, of course, Western Europe[17]. Furthermore, the decline in economic welfare that Western Europe experiences if it bans cereal grain and oilseed imports is changed to a slight gain in this scenario (although recall that Scenario 2 assumes consumers are indifferent to whether a food contains GMOs). Most importantly, the dramatic worsening of resource allocative efficiency in the previous scenario is changed to a slight improvement in this one. Production in the lightly assisted oilseeds sector increases, and production in all other agri-food sectors, which are more heavily distorted in Western Europe, declines. The welfare gains in North America are of similar magnitude in all three scenarios. This result suggests that flexibility in both domestic and foreign markets to respond to these policy and consumer preference changes is considerable. (The cost to North American exporters of seeking out alternative markets to those in Western Europe may be understated in the GTAP model, however.) Given

TABLE 6.3(b). Scenario 2: Selected Economic Welfare Results

	Equivalent Variation (EV) US$ million pa	Decomposition of welfare results, contribution of (US million): Allocative Efficiency Effects	Terms of Trade Effects	Technical Change
North America	2,299	27	-1,327	3,641
Southern Cone	663	71	-303	893
China	804	74	70	669
India	1,277	190	-3	1,092
South Africa	38	3	8	27
Western Europe	-4,334	-4,601	257	0
Sub-Saharan Africa	42	5	38	0
Other high-income	1,371	592	782	0
Other developing and transition econs	1,258	98	523	645
WORLD	3,419	-3,541	0	6,966

Source: Author's GTAP model results.

that the preference shift in this scenario is based on the assumption that non-adopters outside Western Europe cannot guarantee that their exports to this region are GMO-free, Sub-Saharan Africa cannot benefit from the same kind of 'preferential' access the region obtained in the previous scenario, where cereal grains and oilseeds from identifiable GMO-adopting regions were banned completely.

Conclusions and areas for further research

The differences in perception regarding the benefits and risks of the applications of modern biotechnology have already resulted in very different approaches to regulating the cultivation of genetically modified crops and the marketing of their products in Western Europe on the one hand and North America and several developing countries on the other. This in turn has led to the declaration of the Biosafety Protocol, which is essentially a compromise between these views. Moreover, it is yet to be seen whether the stipulations of this Protocol are consistent with WTO rules when carried out in practice. The empirical analysis performed here shows that the most extreme use of its provisions, namely an import ban on GM crops, would have a substantial adverse impact on world production and trade flows not only for the GM-potential crops but also for other related downstream food products. To the extent that some developing and other countries do not adopt GM crops (by choice or otherwise) and they can verify this at the Western European borders, our results suggest it is possible they could gain from retaining access to the GMO-free markets when others are excluded. If European consumers are not worried about GMOs, it is at their expense that non-adopting foreign producers gain. If those consumers *are* worried, on the other hand, the results of the market-based partial preference shift experiment (Scenario 3) suggest that letting consumers express that preference through the market gives markedly greater welfare gains for net-exporting, GM-adopting developing countries. The results also suggest, however, that developing countries that cannot or choose not to access GM technology may lose in terms of economic welfare if they cannot guarantee that their exports entering the Western European markets are GMO-free. For these countries, a complete segregation of GMO-inclusive and GMO-free markets may be a way in which they could reap benefits from selling 'conventional' products to GM-critical consumers in industrialised countries. But to do so would require that the relative price premium on these products is sufficient to outweigh the relative decline in productivity and the potentially significant costs of compliance.

Empirical analysis using models such as GTAP will be able to be enhanced once more data become available. The present analysis had to be based on

TABLE 6.4(a). Scenario 3: Partial Shift of Western European Preferences away from GMOs, Production, Price and Trade Effects, % Changes

	North America	Southern Cone	China	India	South Africa	Western Europe	Sub-Saharan Africa
Production							
Cereal grains	1.8	1.3	1.0	0.4	3.4	-2.0	-2.6
Oilseeds	1.0	2.8	1.1	1	0.0	8.7	-1.6
Livestock	0.9	0.0	0.2	0.4	-0.0	-0.4	-0.1
Meat & Dairy	0.6	0.1	0.1	1.3	0.1	-0.2	-0.0
Veg.oils,fats	1.2	5.0	1.4	-0.0	-2.4	-1.1	-1.2
Other foods	0.2	0.2	0.4	1.5	-0.0	-0.2	0.1
Market prices							
Cereal grains	-5.7	-5.6	-5.6	-6.7	-4.9	-0.2	-0.4
Oilseeds	-5.9	-5.6	-5.7	-6.5	-5.2	0.1	-0.3
Livestock	-1.9	-0.4	-0.4	-1.4	-0.2	-0.1	-0.3
Meat & Dairy	-1.1	-0.2	-0.3	-1.0	-0.2	-0.1	-0.2
Veg.oils,fats	-2.6	-3.3	-2.6	-1.0	-0.5	-0.4	-0.2
Other foods	-0.3	-0.2	-0.5	-1.0	-0.1	-0.1	-0.2
*Exports**							
Cereal grains	6.6	9.7	13.9	34.1	6.5	-29.7	-24.1
Oilseeds	1.4	-4.5	2.1	14.1	-1.9	-41.5	-32.4
Livestock	9.8	-0.9	-3.0	10.0	-1.5	-1.8	-1.2
Meat & Dairy	5.3	-0.4	-0.8	6.0	0.7	-0.7	0.1
Veg.oils,fats	6.7	15.8	5.5	-4.0	-7.6	-5.8	-4.9
Other foods	0.4	0.4	1.7	7.6	-0.4	-0.7	0.1
*Imports**							
Cereal grains	-1.7	-4.8	-3.9	-20.4	-1.7	-23.6	11.5
Oilseeds	-2.9	-9.6	-0.7	-7.4	-0.9	-17.7	17.3
Livestock	-2.3	1.1	0.8	-5.3	1.7	0.4	0.2
Meat & Dairy	-2.1	0.1	0.8	-1.7	0.3	-0.1	-0.0
Veg.oils,fats	-4.2	-3.8	-1.5	3.4	3.1	1.5	3.4
Other foods	-0.1	-0.2	-0.6	-3	0.2	0.1	-0.1

* Includes intra-regional trade.

Source: Author's GTAP model results.

assumptions about the productivity impact of introducing genetically modified crops in the agricultural production system, in the absence of informed views on the size (and factor-bias, if any) of the probable productivity impact for specific sectors in different countries. We have also overstated the effects in so far as we had to use the more aggregated cereals and oilseeds subsectors because the GTAP data set is not disaggregated to the level of maize and soybean. Although the policy response analyzed here – a Western European ban on imports of identifiable GM-crops – is simply hypothetical, note that even some developing countries (e.g., Sri Lanka) have taken that step, so it is not necessarily an unrealistic policy assumption. This scenario assumed that GM-producing regions could at a negligible extra cost place a "may contain GMOs" label on their products and that this would be sufficient for Western European importers to identify and reject such shipments. However, in the case of non-adopting countries, they presumably would face testing requirements at Western European borders. The costs of complying with those procedures may be non-trivial, especially for smaller and poorer developing countries. However, it is conceivable that providing a better guarantee of food safety and/or providing relevant and desired information about the new technology, e.g. via labeling, may effectively stimulate consumer demand (not considered above).

The GM-inclusive and GM-free distinction in the present empirical analysis is based on country of origin: either a country adopts GM crops, or it does not. Ideally, the model should explicitly describe the markets for GM and GM-free products separately within each national market and in the interna-

TABLE 6.4(b). Scenario 3: Selected Economic Welfare Results

	Equivalent Variation (EV) US$ million pa	Decomposition of welfare results, contribution of (US million): Allocative Efficiency Effects	Terms of Trade Effects	Technical Change
North America	2,554	-100	-1,092	3,726
Southern Cone	785	109	-246	917
China	834	106	69	672
India	1,267	184	-9	1,093
South Africa	35	4	4	27
Western Europe	715	393	319	0
Sub-Saharan Africa	-5	0	-7	0
Other high-income	1,233	567	674	0
Other developing and transition econs	1,085	164	289	645
WORLD	8,503	1,428	0	7,081

Source: Author's GTAP model results.

tional market. Such a differentiation has not been made in this paper, but it is clearly an important area of future research.[18] In order to plausibly make such a distinction, more information on consumer preferences for the two types of products will be needed by modelers. That would then open up possibilities for in-depth analysis of identity preservation and labeling systems. Finally, there is obviously scope to look at the market effects of GMOs in other product areas. Of particular importance to developing countries are the prospects for vitamin A-supplemented rice and for pest-resistant cotton varieties. Neither of these is as likely to generate food safety and environmental concerns but both offer great prospects for boosting economic welfare in developing countries (Nielsen and Anderson 2000b). Yet there is the distinct possibility that the extreme reactions against GMOs in some rich countries will thwart the production, dissemination and adoption of GM seeds for all products, not just the ones of concern to rich-country consumers. A less-emotive, more-informed debate is clearly called for so that such opportunities to boost the well-being of potentially billions of poor people are not diminished.

NOTES

1. Definitions of genetic engineering vary across countries and regulatory agencies. For the purpose of this paper a broad definition is used, in which a genetically modified organism is one that has been modified through the use of modern biotechnology, such as recombinant DNA techniques. In the following, the terms 'genetically engineered', 'genetically modified' and 'transgenic' will be used as synonyms.
2. Calculations are based on the FAOSTAT statistical database accessible at www.fao.org.
3. Private seed companies holding patents on their transgenic seeds are able to extract monopoly rents through price premiums or technology fees. In the 1990s they exercised that right, although they are now coming under pressure to at least allow farmers to sow in subsequent years their own retained seed.
4. This has been the strategy of the United States since its adoption in the early 1990s for rBST use in dairying.
5. See Nielsen (1999) and Runge and Jackson (1999) for further discussion of the principles and costs of labeling and identity preservation systems.
6. The precautionary principle implies that considerations of human health and the environment rank higher than possible economic benefits in circumstances where there is uncertainty about the outcome. This principle is already used in certain international agreements concerning chemicals.
7. The WTO TRIPs agreement that protects trade-related intellectual property rights is also very important in the context of the GMO debate, but it is not dealt with in this chapter. For an overview of the TRIPs agreement as it affects developing country agriculture, see e.g. Watal (1999).

8. This product/process distinction came to a head at the WTO in the famous tuna-dolphin case in the early 1990s. The general issue continues to be hotly debated. See, for example, the recent paper by Howse and Regan (2000).

9. The GTAP (Global Trade Analysis Project) model is a multi-regional, static, applied general equilibrium model based on neo-classical microeconomic theory. See Hertel (1997) for a complete documentation.

10. Due to the absence of sufficiently detailed empirical data on the agronomic and hence economic impact of cultivating GM crops, the 5% productivity shock applied here represents an average shock (over both commodities and regions). Changing this shock essentially generates near-linear changes in the effects on prices and quantities.

11. Due to the absence of sufficiently detailed empirical data on the agronomic and hence economic impact of cultivating GM crops, the 5% productivity shock applied here represents an average shock (over both commodities and regions). Changing this shock essentially generates near-linear changes in the effects on prices and quantities.

12. By distinguishing between GMO-inclusive and GMO-free products by country of origin, one concern may be that GM-adopting regions channel their exports to the country or region imposing the import ban (here Western Europe) through third countries that are indifferent as to the content of GMOs and that do not adopt GM technology in their own production systems. The possibility of such transshipments is abstracted from in this analysis. The possibility of such transshipments is abstracted from in this analysis.

13. A technical appendix describing how the exogenous preference shift is introduced into the GTAP model, a method adopted from Nielsen (1999), is available in the longer conference version of this paper (Nielsen and Anderson 2000a).

14. The size of this preference shift is arbitrary, and is simply used to illustrate the possible direction of effects of this type of preference shift as compared with the import ban scenario.

15. When interpreting the trade flow results, recall that the reported results are percentage changes of import and export quantities, flows which are of very different sizes. The model formulation does of course ensure commodity-by-commodity global trade balance, i.e. what is exported must be imported. For data on the production, consumption and trade structures of the various regions in our GTAP model, see Appendix 1 of Nielsen and Anderson (2000a).

16. Of course in so far as Western Europeans value a ban on GM products in their domestic markets, that would offset the welfare loss reported above.

17. Note that the measure of welfare used here is based on efficiency-price considerations. It does not explicitly capture the assumed fact that some consumers and firms in non-adopting Western Europe now prefer domestically produced cereal grains and oilseeds.

18. A first attempt at doing that is presented in Nielsen, Robinson and Thierfelder (2000).

REFERENCES

Buckwell, A., G. Brookes and D. Bradley. 1999. *Economics of Identity Preservation for Genetically Modified Crops*. With contributions from P. Barfoot, S. Tangermann and J. Blom. Final report of a study for Food Biotechnology Communications Initiative (FBCI).

European Commission. 1999. The EU Approach to the Millennium Round: Communication from the Commission to the Council and to the European Parliament. *Note for the 133 Committee*. MD: 400/99.

Food and Drug Administration 1995. *FDA's Policy for Foods Developed by Biotechnology. Emerging Technologies – Biotechnology.* Washington, D.C.: FDA.

Hertel, T.W. 1997. *Global Trade Analysis: Modeling and Applications,* ed. Cambridge and New York: Cambridge University Press.

Howse, R. and D. Regan. 2000. *'The Product/Process Distinction -- An Illusory Basis for Disciplining Unilateralism in Trade Policy.'* mimeo, University of Michigan Law School, Ann Arbor, January.

James, C. 1997 *Global Status of Transgenic Crops in 1997.* ISAAA Briefs No.5. Ithaca, NY: International Service for the Acquisition of Agri-biotech Applications.

————. 1998. *Global Review of Commercialized Transgenic Crops: 1998.* ISAAA Briefs No.8. Ithaca, NY: International Service for the Acquisition of Agri-biotech Applications.

————. 1999. *Global Status of Commercialized Transgenic Crops: 1999.* ISAAA Briefs No.12: Preview. Ithaca, NY: International Service for the Acquisition of Agri-biotech Applications.

James, C. and A. Krattiger. 1999. The Role of the Private Sector. Brief 4 of 10 in *Biotechnology for Developing/Country Agriculture: Problems and Opportunities Focus 2: A 2020 Vision for Food, Agriculture, and the Environment,* ed. Persley, G. J. Washington, DC: International Food Policy Research Institute.

Kelch, D. R., M. Simone and M. L. Madell. 1998. *Biotechnology in Agriculture Confronts Agreements in the WTO. Agriculture in the WTO.* WRS-98-4/December. Situation and Outlook Series. International Agriculture and Trade Reports 34-35. Washington, DC: USDA.

McDougall, R.A., A. Elbehri and T.P. Truong, eds. 1998. *Global Trade, Assistance, and Protection: The GTAP 4 Data Base.* Center for Global Trade Analysis. West Lafayette, IN: Purdue University.

Nelson, G.C., and others. 1999. *The Economics and Politics of Genetically Modified Organisms: Implications for WTO 2000.* Bulletin 809, November. College of Agricultural, Consumer and Environmental Sciences. University of Illinois at Urbana-Champaign.

Nielsen, C.P. 1999. *Økonomiske virkninger af landbrugets anvendelse af genteknologi: Produktion, forbrug og international handel* [Economic effects of applying genetic engineering in agriculture: Production, consumption and international trade] with an English summary. Report No. 110. Copenhagen: Danish Institute of Agricultural and Fisheries Economics (SJFI).

Nielsen, C.P. and K. Anderson. 2000a. *GMOs, Trade Policy, and Welfare in Rich and Poor Countries*. Paper presented at the World Bank Workshop on Standards, Regulation and Trade, Washington, D.C., 27 April 2000 (downloadable as CIES Discussion Paper 0021 at www.adelaide.edu.au/cies).

———. 2000b. *Global Market Effects of Alternative European Responses to GMOs*. Policy Discussion Paper No. 0032. Centre for International Economic Studies, University of Adelaide, Australia.

Nielsen, C.P., S. Robinson and K. Thierfelder. 2000. *Genetic Engineering and Trade: Panacea or Dilemma for Developing Countries?* Paper presented at the Third Annual Conference on Global Economic Analysis 28-30 June. Australia: Monash University.

Organisation for Economic Co-operation and Development. 1998. *Trade and Regulatory Issues and Biotechnology*. Paper prepared by Canada for an OECD Workshop on Emerging Issues in Agriculture. COM/AGR/CA/TD/TC/WS(98)111, Paris: OECD.

———. 1999. *Modern Biotechnology and Agricultural Markets: A Discussion of Selected Issues and the Impact on Supply and Markets*. Directorate for Food, Agriculture and Fisheries. Committee for Agriculture. AGR/CA/APM/CFS/MD(2000)2, Paris: OECD.

Runge, C., F. and L. A. Jackson. 1999. *Labeling, Trade and Genetically Modified Organisms (GMOs): A Proposed Solution*. Working Paper WP99-4, November. Center for International Food and Agricultural Policy. St. Paul, MN: University of Minnesota.

Sykes, A. O. 1999. *Regulatory Protectionism and the Law of International Trade*. University of Chicago Law Review, Winter 66(1): 1-46.

Tietje, C. 1997. Voluntary Eco-Labelling Programmes and Questions of State Responsibility in the WTO/GATT Legal System. *Journal of World Trade* 31(1): 123-158.

United Nations Environmental Program. 2000. Cartagena Protocol on Biosafety to the Convention on Biological Diversity. http://www.biodiv.org/biosafe/biosafety-protocol.htm.

United States Department of Agrigulture. 1999. Value-Enhanced Crops: Biotechnology's Next Stage. *Agricultural Outlook* March:18-25. Washington, D.C.: Economic Research Service, USDA.

Watal, J. 1999. *Intellectual Property Rights and Agriculture: Interests of Developing Countries*. Paper presented at the World Bank and WTO Conference on Agriculture and the New Trade Agenda in the WTO 2000 Negotiations 1-2 October. Geneva.

World Trade Organization. 1995. *The Results of the Uruguay Round of Multilateral Trade Negotiations: The Legal Texts*. Geneva: WTO.

———. 1998a. *Understanding the WTO Agreement on Sanitary and Phytosanitary (SPS) Measures*. May. Geneva: WTO.

———. 1998b. Goods: Rules on NTMs. In *WTO: A Training Package. Module 3*. Geneva: WTO.

———. 1998c. EC - *Measures Concerning Meat and Meat Products (Hormones)*. Report of the Appellate Body, 13 February. WT/DS26/AB/R. WT/DS48/AB/R AB-1997-4. Geneva: WTO.

CHAPTER 7

Quantifying the Trade Impact of Sanitary and Phytosanitary Standards: What is Known and Issues of Importance for Sub-Saharan Africa

T. Ademola Oyejide
E. Olawale Ogunkola
Abiodun S. Bankole

Introduction

Generally, consuming countries require that many domestically produced and imported goods should satisfy certain minimum levels of quality, health and safety standards. These standards are particularly prominent with respect to agricultural, food and health products. Many of these fall under the category of sanitary and phytosanitary (SPS) measures. SPS measures refer to policies designed by consuming countries to protect human or animal life from risks arising from additives, contaminants, toxins or disease-causing organisms in food; protect human life from diseases carried by plants and animals; protect animal and plant life from pests, diseases or disease-causing organisms; and to protect an importing country from the entry, establishment or spread of pests. Other measures have the primary objective of providing information about the characteristics of the products (and thus permit consumers to protect their health and safety). These devices are often referred to as technical standards. They cover a wider range of products, including food, medicines, household appliances, travel goods, mechanical and electrical instruments, and so on. Thus, SPS measures are a subset of the larger class of administrative barriers to the exchange of health-related and safety-related products in national and international markets. They are particularly distinguished by their agricultural and food safety focus (Hooker and Caswell, 1999).

The evolution of SPS measures is traceable to GATT rules, especially Article XX (b), which allows countries to introduce measures to protect human, animal or plant life or health. A further development of this provision is the full-fledged Uruguay Round Agreement on the Application of Sanitary and

Phytosanitary Measures, which contains the definition of SPS measures specified above. The SPS Agreement can be viewed as an attempt to discipline the "seemingly never-ending demand for further regulation of the food system to protect public health" (Henson and Heasman, 1998). While the Agreement confirms the legitimate right of countries to use SPS measures to provide the level of health protection they consider appropriate, it also aims at ensuring that this right is not abused for protectionist purposes and that its exercise does not have unnecessary negative effects on international trade.

The Agreement includes several provisions directed at preventing abuse of rights to use SPS measures (UNCTAD, 1999). First, it encourages countries to base their SPS measures on established international standards, guidelines, and recommendations. It also provides a loophole. Specifically, countries may introduce SPS measures that result in a higher level of protection than would be yielded by those based on international standards if there is scientific justification for the departure, or if the country determines on the basis of appropriate risk assessments, that a higher level of protection would be necessary. Second, the SPS Agreement seeks to promote the international harmonization of SPS measures by encouraging importing countries to accept the SPS measures of exporting countries as equivalent, if they achieve the same level of protection. Harmonization is facilitated further in this respect when countries enter into mutual recognition agreements (MRAs). Third, in deciding among alternative SPS measures that provide the same level of food safety or animal and plant health, countries are encouraged to apply those standards that least restrict trade, provided they are technically feasible. Finally, the Agreement confines the legitimate use of SPS measures to only those necessary for the protection of food safety and animal and plant life and forbids unjustified discrimination whether in favor of domestic producers or among foreign suppliers.

SPS measures can take several different forms. Henson and Heasman (1998) offer the following three broad categories. First are information measures that restrict the behavior of suppliers only to the extent that they are required to disclose specified facts about their products. Under the second broad group are SPS measures that impose prior approval, certifying that their products have met some pre-specified safety criteria before they can be released onto the market. The third class of SPS measures consists of those that allow suppliers to sell products without any prior official approval, but imply that an offense is committed if the products fail to meet certain minimum safety standards.

This form of SPS measure is, in some respects, the least complicated since it does not prescribe any specific safety standards for either the products or the process through which they are produced. It only imposes criminal liability for pre-specified harmful consequences that may result from the use of the product. Compared to this situation, SPS measures incorporating performance standards expect the product to satisfy specified levels of safety, but do not prescribe its production method or process. Finally, some SPS measures incorporate specification standards, which consist of both product and process standards. Thus, they mandate what level of safety the product must achieve and how the product is to be produced, including which ingredients it should or should not contain and the environment of the production process.

Clearly, the impact of a specific SPS measure would depend on the safety level or quality standard specified, as well as the form of the regulatory mechanism (such as product, process or performance standards) through which the specified targets are meant to be achieved (Hooker & Caswell, 1999). This suggests that SPS measures can be highly product and firm-specific, and hence their impact may be difficult to estimate and to aggregate over product groups and firms. At the analytical level, Thilmany and Barret (1997) argue that SPS measures may increase effective demand for the products to which they apply, to the extent that their use relieves consumers' concerns about the quality and safety of such products. But SPS measures can also act as a trade quota. Hence, the effect of SPS measures on the volume of trade is analytically ambiguous.

In this chapter, we identify and discuss what is known about the effects of SPS measures on trade in the products to which they apply in the African context. We consider the extent to which these measures may influence the external market access of Africa's agricultural and food products, and also whether and how these effects could be quantified. The rest of the chapter is structured as follows. In Section 2, we review and assess the practical significance of SPS standards for African agriculture and agricultural trade. The section begins by demonstrating the importance of agriculture and agricultural trade in African economies. We then explain the performance of agricultural and food trade, focusing specifically on factors influencing external market access and the roles of SPS measures in Africa's key foreign markets. Finally, we review SPS measures in African countries and assess the nature of these measures in relation to international norms.

In Section 3, we discuss the available concepts and models for assessing the trade effects of SPS measures. We provide basic analytical propositions about the effects of SPS requirements on production and trade at the firm level, through the imposition of various cost barriers. We also relate the relevant theoretical exposition to African agricultural and food trade using examples from African countries. The section further reviews the available partial and general equilibrium methods and models for quantifying the trade effects of SPS measures in addition to macro-level infrastructure and other associated costs of meeting SPS requirements in African countries. Included is a discussion of the extent to which SPS measures are based on international standards, which basically reflect developed-country norms and practices, and the compliance, which requires African countries to upgrade production, quality control facilities, and laboratories for testing and certification. We conclude in Section 4, by suggesting practical approaches for gathering data and quantifying the trade impacts of SPS measures at both the micro and macro levels, explaining why these tasks may prove useful to African government policy makers and exporters.

SPS Measures and African Agriculture

We review the practical significance of SPS for agriculture and processed agricultural and food exports from Sub-Saharan Africa (SSA). We also discuss the sector's performance and factors influencing its development, focusing on the role of SPS measures in Africa's key foreign markets and the extent to which these measures pose a threat. An important aspect is the tendency for SPS measures imposed by SSA governments to limit export market expansion. This point is brought out from a review of the nature of standards imposed by African governments. The review analyzes whether, and the extent to which, these standards differ from international standards.

Agriculture and the processed agricultural and food sector are of great importance in Africa. Africa's economic growth and development are intrinsically linked to development of its agricultural sector, as the region is highly dependent on this sector for income, employment and export earnings. Agriculture contributes about 35% of the region's gross national product (GNP), serves as the primary income for as much as two-thirds of the population, and accounts for up to 40% of total exports. The sector also supplies factor inputs to the region's fledgling manufacturing sector. In addition, export earnings from the sector finance importation of capital goods necessary for the socio-economic development of the region.

Trends in Agricultural, Processed Agricultural, Food Production and Exports

Trends in production and exports indicate that agriculture has not constituted an engine of growth for the SSA region. Indeed, until recently, various indices revealed downward trends in production and exports. Specifically, the growth performance of African agricultural production shows a sustained recovery in the first half of the 1990s, which recorded an average annual growth rate in excess of three percent. However, the rate of the recovery slowed in the last half of the 1990s, with the average annual growth rate falling below three percent as the sector tipped toward recession. In comparative terms, the average growth performance of SSA agricultural production compared favorably with those in Latin America and the Caribbean (LAC) and Asian regions in the 1990s. However, its performance in terms of regional shares of world merchandise exports lagged behind those of Latin America and Asia, as shown in Table 7.1. This implies, as noted by Oyejide (1999), that the apparent recovery of SSA countries' agricultural production is yet to be translated into rapid and sustainable growth of the sector. Overall, African agriculture has not performed as well as those of the other developing regions. A probable result is that African economies have failed to achieve the kind of structural transformation underlying the decline in the relative significance of agriculture in the more rapidly growing developing economies of other regions.

The unsatisfactory relative performance of Africa's merchandise exports may be explained by the fact that African agricultural exports dominated the total export of the region. Indeed, nine commodities, namely bananas, cocoa, coffee, cotton, groundnuts, rubber, sugar, tea and tobacco contributed 71.2% of total agricultural exports of the region in 1990-97. The two dominant commodity exports, cocoa and coffee, have respectively 18.5% and 14.7% export shares (Townsend, 1999). Cocoa, coffee and groundnuts are on a downward trend while tobacco, tea, bananas, rubber, cotton and sugar have grown in importance. The three leading crops in the 1970s, cocoa, coffee and cotton, which accounted for 54 percent of agricultural exports then, contributed about 45% in the 1990s.

TABLE 7.1. SSA's Performance Indicators (1990-97)

	Agricultural GDP Average Annual Growth (%)	Share of World Merchandise Exports (%)
	1990-97	1990-95
Asia	3.9	24.2
Latin American and Caribbean	3.95	4.5
Africa	3.9	2.6

Source: World Bank World Development Report 2000

Table 7.2 shows the growth performance of major agricultural crops between 1970 and 1997. The table clearly shows that the region has lost market share in all these crops except tea and tobacco, which recorded marginal increases in average annual growth rate of 1.3 percent and 1.7 percent, respectively. Groundnuts recorded the highest market-share loss, at ten percent. Cocoa, coffee and bananas also recorded substantial losses in market share.

The analysis so far focused primarily on Africa's traditional agricultural production and exports. There is, however, an emerging trend with respect to processed agricultural and food production and exports, which signifies the critical role of this sector to SSA. Production of this class of products gained substantial momentum in the 1990s. According to UNIDO (2000), SSA's manufacturing industry is dominated by the processing of raw materials for domestic demand and exports. In particular, while food, beverages and tobacco products contributed 41 percent manufacturing value added (MVA), food processing alone accounted for 23 percent in 1996, making it the largest single sector by this measure. Growth of the processed agricultural and food sector slowed in the 1980s but gained momentum in the 1990s. The share of the sector in total MVA among SSA countries rose from 25 percent in 1980, to 41 percent in 1996.

There is, in addition, an increasing reliance of SSA on the export of processed food products associated with the growing importance of these products in global trade. The SSA countries increased their share of global food processing MVA from 0.5 percent in 1970 to 0.6 percent in 1996. Indeed, the rate of growth of global food processing MVA was less than that of SSA over the 1970-96 period. Furthermore, the food-processing sector accounted

TABLE 7.2. Share of SSA in World Trade (percent), 1970-1997

Crop	Levels 1970	1997	Averages 1970-79	1980-89	1990-97	Growth rate (1970-97)
Bananas	7	4	6	3	4	-3.3
Cocoa	68	41	59	45	40	-2.0
Coffee	26	12	27	22	14	-3.1
Cotton	17	14	13	11	12	-0.2
Groundnuts	68	7	44	11	5	-10.2
Rubber	8	7	6	6	5	-0.5
Sugar	9	7	8	7	8	-0.2
Tea	13	21	15	15	19	1.3
Tobacco	8	12	8	9	12	1.7

Source: Townsend (1999)

for the largest proportion of employment in SSA's manufacturing in 1996, employing 510,000 workers, or 25 percent of total manufacturing employment. The expansion of employment in the sector is traceable to the sector's export performance, which accounted for the largest share (24 percent) of manufacturing exports from SSA in 1995, amounts to US $1.790 billion.

From the perspective of individual countries, traditional agricultural production and exports still form substantial part of economic activity in some countries. However, the processed-food sector in other countries is booming.[1] In particular, Senegalese exports of processed foods reached 53 percent of exports in 1990, as against agricultural raw materials exports of three percent. Its exports of fruits and vegetables were US $2 million in 1990, while fish became Senegal's most important export commodity, with an export value of US $216 million. Direct and indirect employment in the fisheries sector amounted to 150,000 workers. Food processing, in particular oil milling, sugar refining and fish canning, now accounts for more than 40% of industrial output.

Cameroon's exports of food products were 22 percent of merchandise exports in 1993, compared to cocoa at eight percent, coffee at five percent, and fruits and vegetables at seven percent. The growth rate of cocoa butter exports between 1991 and 1993 reached 59 percent while it was 32 percent for food products overall. In Uganda export growth of food and live animals, fish and fish preparations, cereals and cereal preparation, and fruits and vegetables were estimated at 83 percent, four percent, six percent, and 11 percent, respectively in 1993.

In Zimbabwe, the food-products industry in 1990 accounted for some 16 percent of formal employment in the manufacturing sector and 11 percent of the MVA. The beverages industry in that year comprised four percent of employment and 15 percent of MVA. Its contribution to value added grew substantially during the first half of the 1980s and remained steady until 1990. Zambia's food, beverages and tobacco production account for over 60 percent of value added in manufacturing. In Ghana, exports of fruits, especially pineapples and kolanuts, are non-traditional growth areas. Pineapple exports have increased in line with export growth, mainly to EU markets. By 1990, exports of pineapples had risen to US $3.8 million, compared with US $2.1 million in 1989 and US $0.9 million in 1986.

However, in some countries the agricultural and processed food sectors turned in a gloomy performance. Cote d' Ivoire's exports of food products fell to 64 percent of exports in 1992 from 68 percent in 1985. In Ghana, preservation of fruits and vegetables is still a small industry that is well below one percent of manufacturing output. Also, a relatively small amount of production is exported as cocoa butter and paste. In Nigeria, the exports of agricultural and food products either fell or remained stagnant from 1990 to 1996. In particular, the considerable increase in exports of fish and shrimp from 5,600 tons in 1991 to 36,800 tons in 1995 fell to 16,800 tons in 1996.

Athukorala and Sen (1999) identified several key dimensions of trade in processed food in developing countries. First is the significant increase in the share of processed food in total non-manufactured exports relative to that of the developed countries. This share rose from 30 percent in 1970 to 41 percent in 1994 for all developing countries. The corresponding increase for the developed countries was from 14 percent to 35 percent over the same period. The second dimension can be deduced from Table 7.3, which shows the changing composition of processed food exports from developing countries from 1970 to 1994. In 1970, the top five products (preserved vegetables, sugar and molasses, processed meat products, preserved animal feeds, and vegetable oils) accounted for close to 80 percent of the total value of processed food exports from developing countries. By 1994 there was much less concentration, apart from the sharp increase in the share of processed fish products.

Third, Athukorala and Sen (1998) argued that processed food exports have shown greater dynamism than primary commodity exports due, perhaps, to the growth–inducing features that make them resemble manufactured products. Among these attributes are the significant learning effects associated with the production and marketing of processed food products and their

TABLE 7.3. Composition of Processed Food Exports from Developing Countries (% shares)

Products	1970	1994
Processed Meat Products	10.4	7.8
Dairy Products	0.3	0.8
Flour and Cereals	0.9	0.8
Preserved Fruits	3.4	5.0
Preserved Vegetables	25.7	12.1
Sugar and Molasses	23.7	12.1
Coffee Extracts and Chocolate	2.2	2.5
Preserved Animal Feeds	10.3	8.7
Margarine and Food Preparations	0.6	2.3
Beverages	3.0	3.2
Tobacco Products	4.6	7.0
Animal Oils	7.3	11.5
Vegetable Oils	9.6	12.5

Source: Athukorala and Sen (1998) p. 49.

generally lower import-content, in the context of locally available raw materials, which gives them an advantage over import-substitution manufactured products. One piece of evidence of this greater dynamism as exports is the generally higher growth rates of these products when compared with primary commodity exports over the period from 1970 to 1994. Table 7.4 presents relevant data on a number of African countries. Of the eleven countries in this table, only in Cameroon did processed food exports grow less rapidly than primary commodity exports.

Yeats (1998) offered additional evidence to suggest that African exports of processed food products are showing signs of significant dynamism. He showed, first, that intra-African trade in foods and feeds has become relatively more important than trade in other goods. Second, almost half (in export value) of Africa's most dynamic intra-regional trade products are foods and feeds. These exhibited annual average growth rates of 32 to 79 percent between the 1980s and 1990s. Third, as shown by an analysis of revealed comparative advantage (RCA) indices, many African countries have significant comparative advantage in the production of a wide variety of processed food and animal feed products, including fish products, vegetables, cereals, fruits, animal and vegetable oils, beverages, meat, and animal feeds.

TABLE 7.4. Comparative Growth Rates of Processed Food and Primary Commodity Exports, 1970-1994

	Annual Compound Growth (%) 1970-1994	
	Processed Food Exports	Primary Commodity Exports
Burundi	13.7	5.1
Cameroon	1.7	7.5
Côte d'Ivoire	9.0	7.1
Ghana	5.3	3.3
Kenya	8.3	7.7
Madagascar	5.6	2.5
Nigeria	1.1	-0.1
Sudan	4.6	0.4
Senegal	5.0	3.4
Tanzania	5.0	2.5
Zambia	3.1	-0.8

Source: Athukorala and Sen, (1998) p. 45

Explaining the Performance of African Agricultural and Food Exports

Various explanations have been offered for the relatively poor performance of African agricultural and food exports period from the mid–1970s to the mid–1990s. Some of these have focused on the domestic policy environment as mediated through macroeconomic, trade and associated sectoral policies (Oyejide, 1993). Others have concentrated on poor resource endowments (including poor land quality, endemic livestock and human diseases and the like), problems with being land-locked, civil strife and political unrest (Townsend, 1999).

Explanations have been sought in terms of possible external market-access constraints as well. But these typically have been regarded as basically non-existent or negligible. Thus, Ng and Yeats (1996) argued that Africa's extensive loss of competitiveness, which played the key role in its decline in world trade between the 1960s and the 1990s, was due to two factors. First, Africa experienced declining market shares for its major agricultural export products, which were also diminishing in relative importance in world trade. The second factor was the region's inability to diversify its export base.

Amjadi, Reincke and Yeats (1996) demonstrated that OECD tariff and non-tariff barriers could not have been responsible for Africa's poor general export performance, since this region enjoyed trade preferences while its exports faced a much lower NTB coverage ratio than other developing countries. It is worth noting, however, that Low and Yeats (1995) found that most industrial country NTBs are applied to processed food products. This and related evidence probably led Townsend (1999) to conclude that NTBs, such as strict SPS requirements, constitute a potential constraint to the expansion of African agricultural and food exports.

The growth performance of Africa's processed food sector and its ostensible dynamism face substantial threats from SPS measures and other NTBs imposed by developed countries. For example, Nigerian authorities complained that domestic plants processing shrimp for export to the EU were denied certification by customs authorities there, thereby preventing such exports (Trade Policy Review, Nigeria, 1998). Agricultural and livestock production, food products, animal feeds and other products, beverages, and tobacco manufacturing are economic activities that SPS measures affect in Nigeria.

SPS Measures and African Agricultural and Food Trade

Exports from SSA countries go primarily to the EU market (accounting for about 51 percent of exports) and the United States (about 22 percent). Because most exports from SSA countries enter the EU duty free, changes in

EU NTBs are crucial determinants of improved market access for SSA exports. Developments in other markets such as Japan, are also important for diversification of SSA export markets. Using information from the *Trade Analysis and Information System (TRAINS)* software, Version 5.0 (UNCTAD 1997), we compare the NTBs facing SSA-based commodities in the EU with those in the U.S. and Japanese markets.

Despite the fact that *TRAINS* data showed the United States to have the lowest number of NTBs facing agricultural products in the three markets, the United States also has the highest number of regulations regarding food safety (Trade Policy Review, United States, 1999). The United States had issued 144 notifications to the WTO Committee on Sanitary and Phytosanitary Measures by 1998. These regulations may be characterized as the most up-to-date among the three regions. Standards are developed in the United States on a voluntary basis by a mixture of private and government bodies. U.S. standards are enforced at the border, and imported products that do not comply may be refused entry.

Food safety is ensured through pre-market clearance, mandatory production practices, inspections, and random sampling. There are established standards of tolerance for pesticides, herbicides, and fungicides used on foods and other agricultural products. Meat derived from cattle, sheep, swine, goats, horses, poultry and egg products moving in interstate and foreign commerce are safe and wholesome for consumption are accurately labelled. A system of process controls, known as the Hazard Analysis and Critical Control Points (HACCP), has been adopted for the seafood, meat and poultry industries and applied to imported foods. Meat and poultry products may only be imported from countries and plants approved by U.S. authorities, and are re-inspected on a sample basis as they enter the American market. No African nation is listed among the 36 countries eligible to export meat to the United States. Agricultural commodities must meet U.S. import requirements related to grade, size quality and maturity, and require an inspection certificate to indicate import compliance. Seafood products must be analyzed for chemical and biological hazards, including potential hazards in processing.

According to FAO (1999), the United States has rejected or detained imports from Africa for a number of reasons. They include microbiological contamination (41 percent of rejections), filth (18 percent), labeling (13 percent), mold (six percent), decomposition of matter (three percent), low-acid canned foods, food additives, pesticide residue, and heavy-metal content (less than one percent each). These rejections have high costs in terms of lost product value, freight costs, re-export, or destruction. In an independent survey, the United States ranked third among markets for which SPS requirements were considered to be the most significant impediment to trade, behind Australia and the European Union (Henson et al, 2000). The *TRAINS* data indicates that Japan has the highest concentration of NTBs, while the European Union's NTBs facing agricultural exports from SSA are also high. More importantly for this chapter, NTBs in the EU lack transparency in terms of the purpose for which they were imposed. In contrast, NTBs in Japan are explicit and easy to categorize according to purpose. Lack of clear classification of NTBs according to purpose is an issue in quantifying the impact of SPS measures on trade in general.

Table 7.5(a) shows that the EU imposes 13 types of NTBs on agricultural exports to its territory. The designation "product characteristic (health)" is specifically a sanitary standards requirement, while certain other NTBs, such as non-automatic license and prior authorization, may be interpreted as SPS measures. According to the UNCTAD coding system of trade control measures, there are different categories of these (identified) measures. For instance, quotas are coded as group 6200, which has ten sub-classifications. One of these is 6270, comprising quotas imposed for the purpose of controlling imports of sensitive products. This group is further divided into 6271 (protection of human health), 6272 (protection of animal health and life), 6273 (protection of plant health), 6274 (protection of the environment), 6275 (protection of wildlife), 6276 (control of drug abuse), 6277 (assurance of human safety), 6278 (national security), and 6279 (purposes not stated elsewhere).

Table 7.5(a) reveals that only four of 24 two-digit chapters in the harmonized system (HS) of tariff classifications did not face SPS measures in the EU market in 1996. These groups were malt, starches and gluten (HS 11), gum, resins, etc (HS 13), cocoa (HS 18), and preparation of cereal and starches (HS 19). All other chapters encountered at least one form of SPS measure. Explicit SPS measures were applied to the following HS chapters: live animals (HS 01), meat and edible meat (HS 02), dairy products (HS 04) and processed meat and fish (HS 16). According to UNCTAD, these restraints correspond to a frequency ratio of 25 percent and trade-coverage ratio of eight percent.

Table 7.5(b) shows that NTBs applicable to agricultural imports into the Japanese market were dominated by SPS measures. Further, they were both widespread and explicit. Explicit SPS measures included health requirements, authorization (wildlife), labeling for the purpose of health, inspection requirements, standards for human health, and quotas for sensitive products. Indirectly, technical measures not elsewhere stated, import monopoly, and non-automatic licensing could have been imposed as SPS measures. Coverage for explicit SPS measures was widespread. Of the 24 HS chapters only one, meat and edible meat (HS 02), was not affected by health requirements.

TABLE 7.5(a). Non-Tariff Measures facing SSA Countries' Agricultural Exports to the European Union

	Product Characteristics (health)	Prior authorization	Non-automatic licenses	Agricultural levy	Tariff-quota	Variable components	Retrospective surveillance	Seasonal tariff	Automatic license	Reference prices	Bilateral quotas	Quotas	Import monopoly
01 Live Animals	X	X	X	X									
02 Meat and edible meat	X	X		X	X								
03 Fish and Crustaceans		X				X	X		X				
04 Dairy products	X	X	X	X	X								
05 Product of animal origin		X											
06 Vegetable products		X				X	X						
07 Edible vegetables roots/tubers				X	X	X	X		X	X	X		
08 Edible fruits and nuts			X	X	X		X	X		X	X		
09 Coffee, tea and spices												X	
10 Cereals			X		X					X			
11 Malt: starches and gluten					X								
12 Oilseeds, grains, etc.			X	X		X	X						

TABLE 7.5(a). Non-Tariff Measures facing SSA Countries' Agricultural Exports to the European Union (continued)

	Product Characteristics (health)	Prior authorization	Non-automatic licenses	Agricultural levy	Tariff-quota	Variable components	Retrospective surveillance	Seasonal tariff	Automatic license	Reference prices	Bilateral quotas	Quotas	Import monopoly
13 Gum, resins, etc.													
14 Vegetable plaiting material	X												
15 Animal/Vegetable Fats & Oils		X		X			X						
16 Preparation of meat, fish	X	X	X	X	X								
17 Sugars				X									
18 Cocoa						X							
19 Preparation of cereal, starch						X							
20 Preparation of vegetables/fruits			X	X	X	X	X				X		
21 Miscellaneous edible preparations		X				X						X	
22 Beverages, spirits			X	X		X				X			X
23 Residues from food industries		X	X		X								
24 Tobacco					X								X

Source: UNCTAD (1997), Trade Analysis and Information System (TRAINS) software, version 5.0

TABLE 7.5(b). Non-Tariff Measures facing SSA Countries' Agricultural Exports to Japan

	Health Requirements	Authorization on Wildlife	Variable charges n.e.s.	Labeling (health)	Inspection requirements	Non-automatic license	Global quota	Automatic license	Tariff quota	Sole importing agency	Standards (human health)	Seasonal tariff	Technical measures n.e.s	Import monopoly	Quota sensitive products
01 Live animals	X	X	X												
02 Meat and Edible Meat			X	X	X	X									
03 Fish and Crustaceans	X	X	X	X	X	X	X	X							
04 Dairy products	X	X		X		X			X	X					
05 Product of animal origin	X	X	X	X	X X										
06 Vegetable Products	X														
07 Edible Vegetables root/tubers	X		X					X	X						
08 Edible fruits & nuts	X			X							X	X	X		
09 Coffee, tea & spices	X			X											
10 Cereals	X			X			X		X					X	
11 Malt: starches & gluten	X			X					X						
12 Oilseeds, grains, etc.	X	X						X	X				X		

TABLE 7.5(b). Non-Tariff Measures facing SSA Countries' Agricultural Exports to Japan (continued)

	Health Requirements	Authorization on Wildlife	Variable charges n.e.s.	Labeling (health)	Inspection requirements	Non-automatic license	Global quota	Automatic license	Tariff quota	Sole importing agency	Standards (human health)	Seasonal tariff	Technical measures n.e.s	Import monopoly	Quota sensitive products
13 Gum, resins, etc.	X													X	X
14 Vegetable plaiting material	X														
15 Animal/vegetable fats & oils	X	X				X							X		
16 Preparation of meat, fish	X	X	X			X									
17 Sugars	X								X						
18 Cocoa	X												X		
19 Preparation of cereal, starch	X						X		X						
20 Preparation of vegetables/fruit	X								X						
21 Miscellaneous edible preparations	X	X				X			X					X	
22 Beverages, spirits	X								X					X	
23 Residues from food industries	X	X	X			X									
24 Tobacco	X									X					

Source: UNCTAD (1997), Trade Analysis and Information System (TRAINS) software, Version 5.0

Authorization for the purpose of protecting wildlife was required for nine chapters. Labeling requirements for the purpose of protecting health were required in eight chapters.

Compared with the EU and U.S. markets, there seems to be a disproportionate presence of SPS measures in Japan. Perhaps this situation explains in part the concentration of African agricultural exports to the European Union and the United States and the lack of their diversification to the Japanese market. In addition, it is clear that in each of the three markets, the incidence of SPS measures was higher for processed and semi-processed agricultural products than for raw agricultural products.

A few recent studies have examined problems posed by some SPS measures in importing countries. For instance, Mwega and Muga (1998, pp. 39-40) report that "in January 1998, the European Union banned the importation of fresh fish and fish products from Kenya, Uganda, Tanzania, and Mozambique, ostensibly to safeguard EU consumers from the risk of cholera." This decision was based on the claim that these countries lacked a credible system to safeguard fish products from possible contamination. Similarly, "the EU recently introduced controls that subject imported fruits and vegetable to a ten percent sampling for microbial control."

The study suggests that these and other SPS measures could be particularly harmful as they target products in which the countries concerned have a comparative advantage. A parallel study of Uganda (Kasekende, Asea and Abuka 1998, p. 12) claims that "Uganda's exporters, especially of food related items, frequently have particular difficulty in meeting technical regulations, product standards, and sanitary and phytosanitary measures in their main export markets." A third in this series of studies is Oduro and Yahya (1999), who argue that "in the U.S. market, technical standards and health requirements enforced by the Food and Drug Administration act as a barrier to marine imports." They note, in particular, that the automatic (though temporary) testing of Ghanaian exports of fresh and frozen fish and shrimps to the United States "has influenced the decision of some fish exporters in Ghana not to supply that market." The same study refers to EU Directive 91/493, which requires that fish processing establishments and factory vessels must meet specific standards of hygiene. The enforcement of these standards has had at least two effects. First, "of the 38 fishing and fish processing businesses which are registered for general export, only six were approved for export to the EU market." Second, the requirement "has imposed additional cost on some exporters," one of whom "indicated that facilities had to be upgraded at a cost of $5000 per vessel."

SPS Measures in African Countries[2]

Many African countries have set up institutions whose responsibilities include the preparation and drafting of national standards. Some of these standards bodies are also responsible for testing, certification and applying quality controls on domestically produced and imported goods. In this group of countries are Côte d'Ivoire, Ghana, Guinea, Kenya, Mauritius, Nigeria, Senegal, South Africa, Tanzania, Togo, Uganda, Zambia, and Zimbabwe. Botswana, Cameroon, Lesotho, Namibia, and Swaziland have not established formal standards bodies, but depend either on international standards or on other countries' standards bodies. Most Southern African countries, namely Botswana, Lesotho, Namibia, and Swaziland, rely on the South African Bureau of Standards for the promotion and maintenance of standardisation and product quality in their countries. In Botswana, in particular, the choice of specifications and quality assurance is the sole responsibility of producers. In a few countries, such as Mauritius and Uganda, functions of the standards bodies are comprehensive and include the preparation and modification of standards and codes of practice, testing, calibration and inspection of facilities, and suppliers of technical information.

In some African countries, the application of standards is voluntary, while in others it is obligatory. Most standards with respect to products affecting health or safety are compulsory in Mauritius, Cote d' Ivoire, Zambia, Benin, South Africa, Guinea, and Tanzania. In most of these countries, imported products are classified under a list of compulsory standards and must conform to the countries' respective requirements. All standards and other technical regulations that affect health and safety and the environment are also mandatory in Nigeria. The Nigerian standards specify production processes, quality characteristics and mandatory testing procedures.

Most African countries' measures are based on international standards and to that extent are not radically different from international norms. Benin, Ghana, Guinea, Kenya, Mauritius, Nigeria, Tanzania, Uganda and Zambia have standards based on guidelines issued by the International Organisation for Standardisation (ISO), the Codex Alimentarius Commission, and the African Regional Standards Organisation (ARSO). In many of these countries, no instances of inappropriate international standards have been found. Though this is also true of Zimbabwe and South Africa, there are cases in these countries in which international standards arc considered insufficiently, comprehensive and national standards of other countries are applied. In Zimbabwe, for example, the national standards of the United Kingdom, South Africa or other countries are sometimes adopted. The South African Bureau of Standards (SABS) considers its own standards in conjunction with those of ISO, and demands that suppliers' quality management systems comply with SABS-ISO 9000 requirements. South African standards generally follow international norms, and the country had about 93 percent of the 1,558 ISO 9000 certificates held in Africa in 1995. South Africa also has a mutual recognition agreement with the EU regarding calibration and testing facilities. Where there is no MRA, the supplier must prove the compatibility of its products with South African standards.

There are cases in which African standards institutions discriminate against imported products from different origins. For instance, Zimbabwe allows live animals and infectious items to be transported through its territory by air or rail, except when the animals and infectious items originated from, or passed through Kenya, Tanzania, Malawi, Mozambique, and Zambia. Imports of infectious items from the United Kingdom, Republic of Ireland, the United States, Canada, Australia, and New Zealand, or if accompanied by necessary documentation regarding disease, from any other country, may be imported without permit. In South Africa, compulsory standards apply equally to products from all sources, but the importation of live animals susceptible to "mad cow" disease from all countries infected with Bovine Spongiform Encephalopathy (BSE) has been suspended since March 1996. Importation of beef from the UK is also banned. While internationally based national standards in Guinea apply to products for the local market, imported products must comply with approved national standards. However, Ghana and Tanzania do not discriminate between imported and domestically produced goods in applying standards. Tanzania adopts any existing international standard or technical regulation in totality, especially European norms. However, domestic and imported products must meet Tanzanian standards of conformity. Ghana prohibited imports of plants from specified countries through 1992, including maize from the United States, shallots from Togo, and potatoes from Scotland and the United States. Imports of tobacco seeds and cocoa are prohibited from all sources except for research purposes. Zambia subjects all imports of live animals, plants and seeds to quarantine regulations, irrespective of their country of origin. Benin requires a permit for all imports of fresh or chilled fruit and vegetables except from West Africa and Europe, whose goods may enter with a phytosanitary certificate from the country of origin or a transhipment certificate based on the model provided by the International Plant Protection Convention of Rome.

Recognition of testing procedures of exporting countries is another issue in the African application of SPS and other technical regulations. Recognized testing procedures and results conducted in the exporting country are usually accepted in Ghana. So far these procedures have ensured sufficient standards and wholesome quality of products, especially food and drugs. Cote d'Ivoire,

Nigeria, Tanzania, Uganda and Zambia recognize tests performed by bodies that have been approved by the authorities of their home country. Cameroon adopts the standards published by ARSO, but does not recognize testing procedures performed in exporting countries, though there had been no instances where international standards have been found inappropriate. In Mauritius, certificates of analysis from recognized foreign institutions are accepted, but the Health Inspectorate asks for samples of imported food to ensure that shipments comply with local food regulations. Tests carried out in the exporting countries are only recognized by Benin's authorities, if they were conducted by institutions that have been approved by the exporting country, and appear on the official list of entities declared to the Directorate of Food and Applied Nutrition.

The point at which inspections of imports are made is important when considering the SPS issues to the extent that delays, which may result in spoilage and losses, occur in the process of testing imported products. Few African countries, in exceptional cases, test products at the point of entry. In Zambia for example, products are only tested at the borders in suspicious cases. South Africa routinely analyzes food products in specialized laboratories in Durban, Pretoria and Stellenbosch to determine their composition, microbiological contamination and pesticide residue levels. Namibia takes samples from shipments and sends them to South Africa for testing. Guinea's certification, like that of the majority of African countries, is done by SGS during pre-shipment inspection of import products. Up to March 1998, imports into Togo were not systematically inspected.

It is possible that a standard imposed by an African country limits its own exports, if it is a requirement that exports meet the standards of importing countries. How much this occurs is not clear, but there are SSA countries that require inspection and certification of exports. Examples include Mauritius, where the plant quarantine service inspects and certifies all imports and exports of agricultural goods, and Zambia, which has standards for the promotion of exports. While South Africa and Ghana have quality standards for exports, Guinean enterprises employ foreign standards to sell on foreign markets. Tanzania does not test exports unless specifically requested by the exporter.

African countries generally require sanitary and phytosanitary certificates, as well as import permits, on imports of animals and animal products, plants and plant products and other products with potential adverse effects on health. In effect, it is mandatory that a certificate of origin be presented in Togo for the import of cattle. Ghana demands that a sanitary certificate from the veterinary department of the country of origin, certifying that the animals are free from disease, be presented for a permit to import live animals and animal products. Such import permits also require animals to have been slaughtered in an approved abattoir, be free from radioactive contaminants, contain specified minimum levels of fat, and be passed in the exporting country as suitable for human consumption. In addition to requiring that a general phytosanitary certificate accompany imports of plant and plant parts from all countries, importation of any plants (including cassava), and products liable to harbour dangerous diseases or pests need a permit. In Zimbabwe, only agricultural imports are affected by health, sanitary and phytosanitary regulations. While live animals are quarantined for a period of 30 days, and laboratory tests are performed on animal products, a properly completed certificate must accompany imports of plants.

Importers of edible meat, fish and livestock products and feed for other animals must acquire a technical endorsement or a certificate of conformity in Cameroon. Though standardization is at an early stage and imports are admitted with little reference to standards, with exceptions occurring on an *ad hoc* basis, a sanitary certificate is required for the importation of various plant products, including banana plants, cocoa plants, coffee plants, sugar cane, and raw cotton, cotton seeds, cotton plants, and containers holding earth or composts. In Uganda, all imports of live animals, plants and seeds are subject to quarantine regulations and may enter only at designated ports. Entry of any person or animal may be restricted under the Public Health Act, to prevent the introduction of infectious diseases, while certain requirements or conditions (medical examinations, detention, quarantine, disinfection or isolation) may be imposed. Cote d'Ivoire demands a phytosanitary certificate for all imports of plant products and levies a health inspection tax on imports of meat, fish, dairy products, and preserved food. Zambia requires a sanitary certificate from the exporting country as a prerequisite to the issuance of a veterinary permit, while a phytosanitary certificate is required when importing grains. Imports of maize and tobacco (both currently under import prohibitions) and beef into Nigeria require particular sanitary and phytosanitary certificates.

In South Africa, all imports of live animals (including exotic animals) and animal products, plants, pathogens, insects, growth medium, any infectious items, honey, bees and beeswax used as apiary equipment, and genetically manipulated organisms, are subject to import permits. Other agricultural products subject to import control include seeds for planting purposes; fresh and dried fruit, vegetables, flowers and foliage; bark, cork, rattan, bamboo, timber and fibers for manufacturing; vegetable gums and resins; dried herbs, spices, seed and nuts for human and animal consumption and manufacturing; organisms for research, bio-control agents, infected plant material for laboratory analysis; leaf tobacco, tobacco stalks and dust; fresh water fish for aquariums and fresh water crayfish for research. In addition, the use of natural and artificial sweeteners, colorants, and other additives in foodstuffs is regulated, while the sale of foodstuffs containing more than prescribed levels of aflatoxins and other fungus-induced toxins and cereals containing more than specified level of datura seed are prohibited. Namibia subjects all grain imports to phytosanitary inspection and strictly controls the import of live animals by imposing quarantine conditions where necessary. Meat from infected zones can only be imported into the country's free zone if it is deboned or canned, while chicken requires a certificate of freedom from salmonella. In Tanzania, importers of animals, fish, bees, and animal products must receive an import permit and an international zoosanitary certificate.

Prohibitions constitute the main sanctions imposed by African countries where imports do not conform to standards and where imported animals, plants, and their products are not certified as complying with regulatory standards. Examples of countries that explicitly apply such sanctions include Ghana, Senegal, Mauritius, Nigeria, South Africa, Tanzania, and Swaziland. In such countries as Senegal, Nigeria, South Africa and Tanzania, quarantine options are open to importers, whose products do not conform to standards, for a period. South Africa and Zimbabwe have import requirements in terms of veterinary certification and a quarantine period of 30 days for live animals and birds. Kenya destroys or reships, at owners' costs, plant materials which arrive in the country without authority and appropriate documents.

Assessing the Trade Effects of SPS Measures in Africa

A Review of Concepts, Models and Methods

The number of SPS measures dealing with food-related health and safety concerns seems to have multiplied in the developed countries. The growth of international trade in processed and semi-processed food products has increased the prominence of these SPS measures as regulatory trade barriers. Correspondingly, it becomes more important to examine the economic impact of SPS measures on traded products. Key questions in this context relate to the cost of producing an export-quality product and the issue of who will bear the costs of adjusting production processes to meet the SPS standards of importing countries.

Despite the significance of these questions, particularly for the developing countries whose exports are an increasing proportion of processed food products, there is little systematic empirical study of their trade impacts (Thilmany & Barrett 1997, Baldwin 1999). At the conceptual level, an attempt to quantify the trade effects of SPS measures may begin from the analysis and estimation of the firm-specific costs of satisfying the requirements of a specific regulation or standard. Such costs include modification of production processes, testing and certification procedures, and delays in demonstrating compliance. There are also costs imposed by non-compliance with a standard, which failure could influence consumer purchases (Henson and Heasman, 1998). Broadly, therefore, an attempt to determine how SPS measures affect trade flows may be made through an analysis of differences in the costs of compliance.

More specifically, the cost of an SPS measure would include the producers' cost of compliance and the administrative and technical costs incurred by the (usually public) agencies charged with testing and certification and monitoring of producers. Hence, those costs can be decomposed into those incurred at the firm (or micro) level and those incurred at the macro level. Hooker and Caswell (1999) offer an analytical framework for the quantification of the trade impact of SPS measures by focusing on differences in compliance costs that domestic and foreign firms experience in the process of meeting the requirements of such regulatory standards. In this framework, foreign firms may have to bear additional costs in the process of complying with more restrictive national-level SPS regulations of an importing country. These additional costs are an important component of the trade effects of SPS measures.

Both domestic and foreign firms may have to incur sunk costs related to complying with specific SPS regulations. Holleran, et al (1999) argue that these sunk costs are likely to vary by key characteristics, such as product type, the SPS regulatory regime and firm size. In particular, compliance costs may differ across firms and locations due to economies of scale and location-specific cost differences. Small firms could experience disproportionately larger per-unit costs of compliance due to an inability to exploit economies of scale, which arise from in-house quality control facilities or from bulk rates charged by outside testing facilities (Loader and Hobbs, 1999). Empirical support for this proposition comes from Henson and Heasman (1998) who found that unit compliance costs are negatively related to firm size and that "large firms are generally more able to comply with regulations in a manner which yields competitive advantage than small firms."

The location dimension is the primary source of trade effect arising from SPS measures. As Hooker and Caswell (1999) argue, SPS measures are often associated with an asymmetric cost increase for foreign producers. There are at least two sources of this asymmetric cost increase (Holleran, et al 1999). An importing country's SPS measures may specify idiosyncratic production and processing standards that place foreign firms at a competitive disadvantage in comparison with local producers. In addition, the certification process may be "captured" by local producers, imposing undue cost on foreign suppliers. Hooker and Caswell (1999) show that when an SPS regulation asymmetrically increases the compliance costs of foreign producers, it will have a "tariffication effect" of raising equilibrium price, reducing imports, and increasing domestic production, even though no tariff revenue is generated. The reduction in imports results directly from the increased cost of compliance imposed on foreign suppliers.

This compliance-cost approach for quantifying the trade effect of SPS measures may be implemented by estimating a cost function that incorporates output and "quality" dimensions, where the latter element captures the requirements of the specified SPS measure. In Antle (1999), the basic cost function consists of three key components. Total cost is made up of a conventional fixed-cost component, which is independent of both output and quality, a variable-cost component that combines conventional production inputs with some inputs used for achieving the specified SPS standard, and a second variable-cost component that captures SPS compliance costs that do not depend on conventional production inputs. In this framework, cost of compliance is captured by the change in total cost of production that arises from complying with the SPS standard. In principle, "before" and "after" production cost data may be used to estimate the cost of compliance.

Antle (1999) identifies and describes three cost-estimation methods that could be employed. The accounting approach focuses on the consequences of meeting specified regulatory standards for increases in input requirements and costs but does not involve a parametric estimation of the cost function. The economic-engineering method combines detailed engineering data with corresponding input-cost data to construct a quantitative model of the production process. The resulting production function may then be used to derive a parametric cost function. Finally, the econometric method can be used to establish a parametric cost function. It also offers a robust basis for statistically testing both structural and behavioural relationships embedded in the responses of producers to the SPS regulatory regimes.

When analyzed from the perspectives of both the importing and exporting countries, the effects of SPS measures replicate the traditional analysis of tariffs and quotas. One important reason is that the effects can be viewed purely as restricting trade, since their application can lead to either a total or partial ban of the affected product. In the case of SPS-induced total import bans, the good is prohibited from entering the importing country. The effects would be similar to those arising from a prohibitive tariff because the expense involved in meeting conformance requirements would make the post-certification import price higher than the domestic price of the importing country. A partial import ban, on the other hand, is equivalent to applying non-prohibitive conformance requirement, where the good is allowed into the importing country conditional on quarantine inspection, fumigation or isolation (James and Anderson, 1998). Thus, the effects would be similar to those of a non-prohibitive tariff.

Baldwin (1999) analyzed theoretically three other possible effects of SPS or TBT requirements. Specifically, such measures could raise the marginal costs or fixed, one-time market entry costs of exporting firms, as well as introduce delays for the foreign goods. When a measure is imposed the domestic price would differ from the border price by the conformity costs, which would be added to the per-unit cost of importing the product. The effects would include a rise in the domestic price and production of the good in the importing country and a decline in domestic consumption, welfare, and imports. In his Cournot market framework, an SPS measure of a fixed-cost type could lower price in the country imposing the measure, and increase the exports of foreign firms that continue to operate, because of increased competition. The effects of delay-inducing measures would be similar to those arising from a temporary quantitative restriction if the barrier does not have a fixed-cost or variable-cost component. If such components exist, the price of the domestic good would rise above world price, thus inducing a fall in domestic consumption as total imports fall. When delay barriers are combined with fixed-cost barriers, additional fixed costs must be absorbed to convince consumers to switch to the product experiencing pre-entry delays.

For competitive export firms, such as processed food enterprises from SSA, the imposition of SPS measure would raise one-time entry costs. The impact would be similar to that from a fixed tax, generating an increase in the per-unit cost of the export good. Thus, substantial entry or compliance costs are likely to discourage exports from small-scale firms, such as the majority of African exporters. Moreover, African countries are disproportionately affected by SPS measures because of the high degree of sophistication required in testing and monitoring, especially in terms of identifying limits on pesticide residues, heavy metals and other contaminants. These considerations make the SPS issue problematic to African countries to the extent that the measures effectively impede African exports to developed countries. The severity of this impediment was assessed by Mutasa and Nyamandi (1998). Their major finding was that 57 percent of African countries had products rejected during border inspection in developed countries since 1996 for reasons of microbiological contamination and spoilage. The income of Tanzanian fishermen fell by 80% during the second period of restrictions imposed by the EU in 1997 on the imports of fish from certain SSA nations. These restrictions were put in place because salmonella was detected in the fish from tests in 1996 and 1997 (Henson, et al, 2000). A second restraint was a port-of-entry test for Vibrio cholera and Vibrio Parahaemolyticus in 1997.

In addition, African exporters of agricultural and food products were often offered lower prices for their exports by buyers in developed country due to perceived low quality. In Cameroon and Ethiopia for example, exporters of coffee and cocoa were often paid lower prices due to problems affecting quality as identified by European buyers. The fact that the crops were inspected before shipping according to international norms did not alleviate the perceived quality problems.

African exporters are often slow to acquire information about relevant new SPS measures imposed by developed countries. Insufficient financial resources to establish and monitor food controls appear to be a major problem associated with prompt and sustained compliance with SPS measures. Indeed, according to Mutasa and Nyamandi (1998) the most important impediment to food exports associated with SPS standards is inadequate financial resources for food controls. A number of further problems faced by African exporters regarding SPS were identified by Henson, et al (2000). These include access to compliance resources, short compliance periods allowed by developed countries, and the slow reactions of African governments to changes in major export markets. Additional difficulties were incompatible production methods, cargo or airfreight problems, and inadequate access to technical information.

In light of the problems discussed, African policy makers should urgently take steps to alleviate SPS problems. One step would be to harmonize their standards with international norms and trading-partner standards to increase their exports. Harmonization of African standards with international norms is likely to expand Africa's market opportunities since products could be exported to all developed countries that apply those standards. This point is validated by the theoretical analysis of Thilmany and Barret (1997), who incorporated the possibility that the SPS measure could stimulate demand in the importing country by resolving uncertainty about product quality and safety. This process would raise excess demand in the importing country and increase the equilibrium price and trade volume relative to the 'no SPS measure' trade scenario.[3]

Since most African countries based their national standards on international norms, effective implementation of SPS requirements on export goods would facilitate such harmonisation and reduce the impact of regulatory barriers. African governments should therefore move to ensure the acceptability and credibility of their standards authorities. Continued lack of confidence in African standards authorities responsible for testing, inspection, and certification would result in renewed tests in importing countries, and cause undue delays or rejections. Harmonization is justified by the requirement of the SPS Agreement that members should recognize and accept each other's test procedures and SPS measures where they can be demonstrated to be equivalent. A case in point is Zimbabwean horticultural products, which were inspected at the EU border, despite Zimbabwe's confirmation that the products were free of pests and diseases prohibited in the EU. Harmonization of SPS requirements would eliminate this double incidence of tests and inspection by adopting international standards, demonstrating compliance and ensuring transparency and equivalence.

African states need to develop a mechanism capable of demonstrating the equivalence of their SPS measures to those of developed countries. This is necessary to counter the leverage provided in the SPS Agreement, whereby members can adopt measures that achieve a higher level of protection if they are scientifically justified. African countries thus require a scientifically oriented mechanism to investigate new SPS measures of developed countries. This would help increase their capacity to consult with authorities in developed importers or to institute procedures on trade disputes with the Dispute Settlement Body (DSB). Though most African countries have enquiry points in respect of SPS issues, efforts should be made through bilateral arrangements to ensure timely acquisition of information regarding new SPS measures that could threaten export interests. Also, notifications to enquiry points should be required to contain sufficient information to permit deployment of appropriate compliance resources.

African governments should effectively participate in the SPS Agreement. Most African countries became signatories by virtue of implementing the Uruguay Round Agreement. In this respect, African governments should adequately identify, analyze and protect their acquired trade rights under the SPS Agreement. African policy makers therefore need to amplify the special needs of the region in the application of SPS measures by identifying and advocating a phased application of SPS measures regarding the region's products of special interest. Compliance periods, which are the time intervals between notification and application, may need to be extended to obtain adequate compliance resources and for impact assessment.

Cooperation among African governments within a regional or multilateral framework is necessary for the adoption of a common position on SPS issues, provision of alternative funding mechanisms to increase capacity to participate in the SPS Agreement, and solicitation of effective legal, scientific and technical assistance from developed countries. Regional or multilateral cooperation would also ensure that developed countries take full cognizance of the potential impact of SPS measures on African countries' particular interests and circumstances. In addition, a cooperative and consultative framework could facilitate the sharing of experience, responsibilities, expertise and information relating to SPS measures.

Quantifying the Trade Effects of SPS Measures: An African Perspective

If African countries wish to expand access to the growing global market for processed agricultural and food products, they must be prepared to meet the SPS measures imposed by importing countries. The costs of meeting these standards go well beyond the firm-level compliance costs discussed earlier. They also include infrastructural costs, or costs of services provided by public agencies. These macro costs can be substantial, given the often large gap between current African standards for export of processed agricultural and food products and the corresponding levels of internationally recognized standards (Finger and Schuler, 1999).

Several key elements of these costs of compliance with SPS regulation can be identified. These elements correspond to particular features of the SPS Agreement. First, the Agreement provides that, as much as possible, countries should base their SPS measures on standards established or recommended by the appropriate bodies. It also suggests that the active participation of developing country members in these international bodies should be encouraged and facilitated. For the typical African country, effective participation in the activities of these bodies not only would require attendance at their meetings but also, and more importantly, acquisition of the technical capacity to understand and contribute to the process of international standard-setting. Both activities involve costs that may be substantial, especially as the existing level of technical knowledge could require substantial upgrading.

Second, the SPS Agreement permits countries to set their standards above those suggested by the international bodies on the basis of their own risk assessment. Such extra-normal standards can be challenged in the WTO if the exporting country negatively affected has the technical capacity to do so. In many African countries, such a challenge would be difficult to mount due to the absence of the appropriate legal framework and a lack of scientific knowledge, supporting infrastructure and trained personnel to carry out relevant risk assessments. The costs of creating and strengthening the capacity to perform these tasks at a global level of sophistication may not be immediately affordable by many African countries.

Third, the Agreement urges countries to accept the results of conformity assessment procedures carried out in other countries. These procedures involve inspection, testing and certification and are often carried out by public or quasi-public agencies in many African countries. Exports of various types of processed agricultural and food products, particularly fish, from African countries have encountered market-access problems in the developed countries because their conformity-assessment bodies are not regarded as credible. Thus, further costs must be incurred to upgrade the human, technical and infrastructural resources of these bodies.

Finally, the Agreement seeks to promote the global harmonization of SPS measures by encouraging members to enter into mutual recognition agreements (MRAs). This may cover types of tests and testing methods as well as conformity-assessment certification. Costs associated with entering into these MRAs are high. In addition, as Baldwin (1999) argues, "the MRA approach ... require(s) a high level of trust in a nation's governance capacity," which few African countries have been able to generate.

In the aggregate, the macro costs associated with compliance in African countries could be quite significant. In a preliminary effort to establish the broad levels of these costs, Finger and Schuler (1999) suggested that they could amount to a large proportion of the typical African country's annual development budget. Given this kind of estimate, Baldwin (1999) advocates that "industrial nations might directly or indirectly subsidize conformance assessment procedures for products made in developing nations." Note, however, that the SPS Agreement itself recognizes that certain African and other developing countries may experience difficulties in fulfilling their obligations. Some attempts to ameliorate these difficulties exist in a set of provisions that offer technical assistance, especially in the areas of processing technologies, research and training, and infrastructure. The technical advice promised could take the form of consulting, credits, and grants and should apply particularly "where substantial investments are required in order for an exporting developing country Member to fulfill the SPS requirements of an importing Member." These technical-assistance provisions have, however, remained largely a statement of good intentions rather than realized mechanisms for lowering the costs of meeting SPS standards.

Quantifying the Impact of SPS Measures on African Exports: A Multi-Stage Approach

In order to capture fully the trade impacts of SPS measures on African exports of processed agricultural and food products, it is necessary to design a data-gathering methodology that covers both the firm-level and macro costs identified and discussed above. A multi-stage approach would be appropriate. The first stage could cover the identification, in each African country, of the processed agricultural and food products, and a detailed analysis of the production processes involved. A second stage would identify the applicable SPS measures in the major export markets, and analyze their key features. Thus, the analyst would determine the standards each nation specifies, whether these are process or performance standards or both, what conformity assessment procedures are used to enforce the standards, and how the procedures implemented. The third stage would model and estimate firm-level cost functions. This stage would draw upon the results of the earlier phases and would add appropriate cost data, derived from detailed firm-level surveys. The fourth stage would focus on "macro" compliance costs, particularly those relating to conformity-assessment procedures for different products. This final stage should yield cost estimates associated with conformity assessment, product-quality monitoring and control services by public and quasi-public agencies for each product or product group.

Each of the suggested stages would require data collection, estimation and analysis. Data could be gathered from a representative sample of such firms in each product group. Econometric techniques should be used, where the quality and depth of data permit, to estimate the cost functions of firms. From these estimates could be drawn the estimated firm-level compliance costs imposed by the different SPS measures.

The firm-level survey approach to data gathering is widespread in Africa. However, since the quantification of the trade impact of SPS measures would have a detailed focus on production and processing costs and would relate specifically to exporting firms, the survey design would have to be more specific than usual.

Purposeful data gathering is imperative in the attempt to quantify the price, trade volume and welfare effects of SPS measures on exporting firms and countries. The firm-level survey approach could discern changes in production costs, export revenues and profit margins induced by SPS measures. Furthermore, data on export prices and quantities would be required at the macro level to capture aggregate effects. For example, the data-collection task could emphasize how to capture changes in market prices associated with SPS measures.

The quantification and estimation of the costs of SPS measures on African countries is particularly appealing in order to draw attention to the circumstances of Africa in the ongoing SPS debate. Policy makers and exporters need to be made aware of the full range of options available for coping with SPS issues and the opportunities that might be gained from implementation and cooperation.

NOTES

1. See World Trade Organization (various years).

2. Source for this material is WTO (various years).

3. James and Anderson (1998) considered a different case, wherein a disease imported into a country could prompt adoption of more disease-prevention and spraying programs. These activities could raise costs of domestic production, thereby reducing domestic supply and increasing imports.

REFERENCES

Amjadi, A, U. Reincke and A. J. Yeats. 1996. Did External Barriers Cause the Marginalization of Sub-Saharan Africa in World Trade? World Bank Policy Research Working Paper No. 1586. Washington, D.C.: World Bank.

Antle, J.M. 1999. Benefits and Costs of Food Safety Regulation. Food Policy v. 24, 6: 605–124.

Athukorala, P.C. and K. Sen. 1998. Processed Food Exports from Developing Countries: Patterns and Determinants. Food Policy v. 23: 1: 41 – 54.

Baldwin, R.E. 1999. Frictional Trade Barriers, Developing Nations and a Two-Tiered World Trading System. Manuscript. Centre for Economic Policy Research (CEPR) Workshop on *New Issues in the World Trading System*. London, 19-20 February.

FAO (Food and Agricultural Organization). 1999. Importance of Food Quality and Safety for Developing Countries. Committee on World Food Security 25th Session. Rome: FAO.

Finger, J. M. and P. Schuler. 1999. Implementation of Uruguay Commitments: The Development Challenge. WTO/World Bank Conference on *Developing Countries in a Millenium Round*. Geneva, 20 – 21 September.

Henson, S. and M. Heasman 1998. Food Safety Regulation and the Firm: Understanding the Compliance Process. Food Policy, v. 23, 1: 9 – 24.

Henson, S., R. Loader, A. Swinbank, M. Bredahl, and L. Nicole. 2000. Impact of Sanitary and Phytosanitary Measures on Developing Countries. Centre for Food Economics Research, Department of Agricultural and Food Economics. United Kingdom: The University of Reading.

Holleran, E., M. E. Bredahl, and L. Zaibet. 1999. Private Incentives for adopting Food Safety and Quality Assurance. Food Policy, v. 24, 6: 669 – 684.

Hooker, N. H. 1999. Food Safety Regulation and Trade in Food Products. Food Policy, v. 24, 6: 653 – 668.

Hooker, N. H and J. A. Caswell. 1999. A Framework for Evaluating Non-Tariff Barriers to Trade Related to Sanitary and Phytosanitary Regulation. Journal of Agricultural Economics, v. 50, 2: 234 – 246.

James J. and K. Anderson. 1998. On the Need for more Economic Assessment of Quarantine/SPS Policies. The Australian Journal of Agricultural and Resource Economics, v. 42, 4: 425 – 444.

Kasekende, L. A., P. K. Asea, and C. A. Abuka. 1998. Trade Policy, Manufacturing Efficiency and Exports in a Liberalizing Economy. Paper prepared for *African Economic Research Consortium (AERC) Project on Africa and the World Trading System*. Mimeo. Nairobi: AERC.

Loader, R. and J. E. Hobbs. Strategic Responses to Food Safety Regulation. Food Policy, v. 24, 6: 685 – 706.

Low, P. and A. Yeats. 1995. Non-tariff Measures and Developing Countries: Has the Uruguay Round Leveled the Playing Field? The World Economy, v. 18, 1: 51 – 70.

Mutasa M.P. and T. Nyamandi. 1998. Report of the Survey on the Identification of Food Regulations and Standards within the Africa Region Codex Member Countries that Impede Food Trade. Paper presented at a *Workshop on Codex and Harmonization of Food Regulations*, August 1998, Harare.

Mwega, F. M. and K. F. Muga. 1998. Africa and the World Trading System: A Case Study of Kenya. Paper prepared for the *AERC Project on Africa and the World Trading System*. Mimeo. Nairobi: AERC.

Ng, F. and A. Yeats. 1996. Open Economies Work Better! Did Africa's Protectionist Policies Cause Its Marginalization in World Trade? Policy Research Working Paper 1636, Washington, D.C.: World Bank.

Oduro, A. D. and K. Yahya. 1999. The Uruguay Round and Ghana. Paper prepared for the *AERC Project on Africa and the World Trading System*. Mimeo. Nairobi: AERC.

Oyejide, T. A. 1993. Effects of Trade and Macroeconomic Policies on African Agriculture. Chapter 12 in R. M. Bautista and A. Valdes (eds), *The Bias Against Agriculture: Trade and Macroeconomic Policies in Developing Countries*. San Francisco: ICEG/IFPRI, ICS Press.

———. 1997. Africa's Participation in the Post-Uruguay Round World Trading System. PSIO Occasional Paper, Series No. 6. World Trade Organization (WTO): Geneva.

———.1999. African Agriculture in the WTO Framework. Paper prepared for the *AERC Project on Africa and the World Trading System*. Mimeo. Nairobi: AERC.

Thilmany, D. D and C. B. Barret. 1997. Regulatory Barriers in an Integrating World Food Market. Review of Agricultural Economics, v. 19, 1: 91 – 107.

Townsend, R. F. 1999. Agricultural Incentives in Sub-Saharan Africa: Policy Challenges. Technical Paper No. 444. Washington, D.C. :World Bank.

United Nations Conference on Trade and Development (UNCTAD). 1997. Trade Analysis and Information System (TRAINS) software, Version 5.0. New York, NY: United Nations.

———. 1999. Preparing for Future Multilateral Trade Negotiation: Issues and Research Needs from a Development Perspective. New York, NY and Geneva: United Nations.

United Nations Industrial Development Organization (UNIDO). 2000. African Industry 2000: The Challenge of Going Global. Vienna: UNIDO.

World Bank. 1999. World Development Report 1999. New York, NY: Oxford University Press.

World Trade Organization. Various years. Trade Policy Review, Various Country Issues. Geneva: WTO.

Yeats, A. J. 1998. What Can Be Expected from African Regional Trade Arrangements? Some Empirical Evidence. Policy Research Working Paper Series #2004. Washington, D.C.: World Bank.

CHAPTER 8

Measuring Environmental Compliance Costs and Economic Consequences: A Perspective from the U.S.

Francis (Hank) G. Hilton
Arik Levinson

This chapter discusses the possibility of measuring international differences in environmental compliance costs, for the purpose of assessing their effect on international trade and investment, using lessons learned from comparing compliance costs in U.S. states. Two major lessons stand out. The first is the critical importance of measuring standard stringency in more than one year, in order to study the effect of changes in standard stringency on changes in international trade. Cross-section analyses, done at one point in time, are doomed by omitted variable bias and the endogeneity of the standards. Intertemporal comparisons could be generated by planning to conduct surveys over successive years, or perhaps more easily by replicating a previously completed survey. The second lesson is that comparisons of national environmental policies can easily be distracted by lots of issues unrelated to manufacturer costs. If the goal is studying the effect of policies on trade, only those policies directly related to manufacturers' costs need to be included.

For decades now, policymakers and policy analysts have been concerned with the effect of domestic environmental standards on international trade. The concerns come from all sides. Business interests and organized labor in developed countries worry that high environmental compliance costs will cause industry (and jobs) to relocate to developing countries. Free trade advocates express concern that nations may be able to evade international tariff agreements by strategically setting environmental policies (Ederington, 1998; Copeland, 1990). Environmental organizations at one extreme fear that industries will use international trade to avoid strict environmental regulations by relocating the pollution-intensive parts of their manufacturing processes overseas. At the other extreme, however, environmentalists have also expressed concern that linked trade agreements will prevent nations from setting their desired levels of environmental protection.

Many of these concerns have been reflected in policy proposals. Twenty-years ago, CBS News Anchor Walter Cronkite wrote a letter to the *New York Times* arguing that the United States should act unilaterally to "protect both American industry and the environment by barring products from any country that does not enforce pollution standards as strict as our own."[1] In 1991, U.S. Senator Boren (D-OK) sponsored the International Pollution Deterrence Act, which, had it been enacted, would have imposed a tariff on imports of products from countries without "effective pollution controls." The proposed tariff was to be equal to the costs foreign producers would have had to incur in order to comply with U.S. environmental standards.

These policy proposals, and their accompanying concerns, are based on several unverified assumptions. First, they presume that international trade, and international industry locations, are sensitive to differences in countries' environmental compliance costs. Second, they presume that those compliance costs are higher in industrialized nations than in developing countries. And third, in the case of Senator Boren's bill, they presume that such compliance cost differences are measurable.

This chapter discusses the possibility of measuring international differences in environmental compliance costs, for the purpose of assessing their effect on international trade and investment, using lessons learned from comparing compliance costs in U.S. states. The next section discusses the practical and theoretical problems inherent in such an exercise. Section 2 describes an attempt to measure compliance costs and their effects on international trade using data from U.S. states. Section 3 provides a summary of previous attempts to measure regulatory stringency differences internationally, and, drawing on lessons learned from the U.S. experience, and from past international surveys, suggests general strategies that might be implemented in any future international survey of compliance costs.

Problems Measuring Compliance Costs

The first item to clarify is the *goal* of the measurement task. One can easily imagine a research project designed to measure differences across countries in environmental *quality*, or in citizens' valuation of the environment, relative to other goods, or in protection of endangered species, depletion of natural resources, automotive emissions control, or recycling efforts. For example, in 1991 Island Press published the widely cited *Green index: A State-by-State Guide to the Nation's Environmental Health* (Hall and Kerr, 1991). The index essentially sums the number of statutes each U.S. state had from a list of fifty common environmental laws. These include statutes pertaining to consumer recycling programs, parks and recreation, agriculture, and transportation. The majority of these statutes are unlikely to be related to manufacturing costs, and would not be useful in assessing the effect of environmental regulations on interstate or international commerce.

However, the goal of this chapter is to measure the effect of environmental regulations on international trade. Consequently, we focus on those regulations that affect manufacturers' compliance costs. In fact, for the purpose of measuring the effect of compliance costs on trade, we do not care whether those costs result from government imposed regulations, or from some other source. Environmental compliance costs may be driven by geographical features of the country, by citizens' preferences and citizens' activism, by differences in factor cost of inputs to pollution abatement, or by self-imposed industry norms. Regardless of their source, we are interested in measuring compliance costs that put manufacturing industries at a relative disadvantage with respect to manufacturers in other countries or other states.

In the end, we would like to be able to answer the question "how much higher would the environmental compliance costs be for a given manufacturer if it were to locate in country *Y* rather than in country *X* ?" This turns out to be a difficult concept, both practically and in theory. First, in practice, it involves surveying many manufacturing plants of different sizes, in different industries, and of different vintages. Even with the resources to conduct such a survey, however, compliance costs will be difficult for manufacturers to assess. Many costs associated with pollution abatement also generate a certain amount of cost savings – so-called "cost offsets." And many efficiency enhancing investments that firms might have made in the absence of pollution regulations may be environmentally friendly. Segregating the components of investment or operating costs that have purely environmental purposes will be difficult.

Because of the difficulty acquiring such information, in theory and in practice, many studies have relied on proxies for these cost differences. The most common approach is to assume that developing countries have lower standards and compliance costs than industrialized countries. This assumption runs through the 1992 World Bank volume, edited by Patrick Low. Its most recent manifestation is a working paper by Matthew Kahn (2000). Kahn finds little evidence that the U.S. has increased imports of pollution-intensive manufactured goods relative to clean goods, or that imports of pollution-intensive goods from poorer nations have increased from 1972 to 1992.

Alternative proxies for compliance costs that have been used by researchers attempting to measure their effect on international trade include ambient environmental standards, observed emissions by firms, ambient environmental quality, and specific provisions of national laws. These proxies, by virtue of their being indirect measures of compliance costs, all suffer from one or more biases and problems of interpretation. A nation's ambient environmental standards may be more or less difficult to comply with depending on its geography. Similarly, ambient environmental quality may be a function more of geography than of regulatory stringency. For example, in the U.S., the Los Angeles metropolitan area has the worst air quality, the most stringent regulations, and the highest associated compliance costs (Berman and Bui, 1999). Specific provisions of national environmental laws are not necessarily better proxies for compliance costs, as those laws may or may not be enforced strictly.

In an ideal world, we would have a direct measure of compliance costs for a variety of industries, firm sizes, and years. This would enable us to compare compliance costs for similar manufacturers in different countries. While we recognize that such data would be difficult to gather, both conceptually and practically, and that most studies make do with a rough proxy for the hypothetical ideal compliance cost data, it is worth describing the ideal before resorting to the proxy.

Furthermore, it turns out that even with the ideal data, measuring the effect of compliance costs on trade requires overcoming several enormous pitfalls. These include the endogeneity of policies, capital vintage effects, and industrial composition effects. Below we describe each of these three problems.

Problem #1: Endogenous Policies

The most important obstacle to measuring the effect of compliance costs on trade is that regulatory stringency and compliance costs themselves may be driven by the trade. For example, a country endowed with an attribute desirable to a polluting industry: a great natural harbor, an abundant key resource, inexpensive energy, etc. Such a country will be likely to attract export-oriented investment in that polluting industry. If the resulting pollution is bad enough, the country will be induced to regulate stringently the pollution emanating from the industry, increasing that industry's environmental compliance costs. In this example, both investment and compliance costs will be *positively* correlated with the presence of the desirable attribute, inducing a spurious positive correlation between investment and environmental stringency.

As another example, suppose that some country has a tendency to favor polluting industries, perhaps because those industries are particularly important to the country's economy, or because those industries have long histories in the country, or because the leadership of the country is "pro-business" in some sense. Such countries, in addition to providing lax environmental regulations, may also provide tax relief or other subsidies to lure foreign investment. In that case, because the low environmental compliance costs are correlated with other inducements, compliance costs and investment will be *negatively* correlated, though not necessarily due to the environmental compliance costs.

In both examples, some unobserved characteristics of countries, key attributes or pro-business climates, were correlated with both compliance costs and investment, inducing a spurious correlation between the two. If the estimation strategy used to measure the effect of compliance costs on investment fails to account for this spurious correlation, then it will impart an omitted variable bias on the predicted effect of regulatory compliance costs on investment.

By far the easiest way of accounting for countries' unobserved characteristics that do not change over time is to collect a panel of data and examine how *changes* in compliance costs affect *changes* in investment. In other words, if we look at investment in various countries as a function of their compliance costs *in one year*, we will be unable to discern whether investment and compliance costs are high due to the presence of natural assets desirable to polluting industries, or whether more investment flows to low-cost countries merely because of other non-environmental inducements provided by high-cost countries. However, if we collect data from two different years, and examine how investment or trade changed in countries where compliance costs grew more, relative to how investment or trade changed in countries where compliance costs grew less, assuming that key resources and pro-business sentiments remained unchanged, that will provide a closer approximation to the true effect of compliance costs on investment.

Unfortunately, *all* of the existing surveys of environmental regulations across countries are single cross-sections. Such surveys tend to be funded once, for a specific subset of countries, with a specific set of questions, and then abandoned. In many ways, research on this topic would be aided most if the next international organization to fund a survey of environmental regulations, rather than designing a new survey with new questions for yet another subset of countries, would instead merely repeat some of the surveys conducted in the past, using the same countries and methodologies as was used before. Later, in section 3 of this chapter, we describe some of those previous international comparisons of regulatory stringency.

Problem #2: Capital Vintage

The second conceptual problem facing measurement of compliance costs involves differences across countries in the age of their manufacturing capital stock. Any survey of manufacturers will, by necessity, measure the compliance costs for *existing* manufacturers. For a variety of reasons, this may be starkly different than the compliance costs for *new* investment. In the U.S., for example, many state laws, and the federal New Source Performance Standards, impose more costly regulations on new sources of pollution than on existing sources. These laws effectively "grandfather" existing sources from the new, more stringent standards.

This grandfathering of existing sources of pollution raises several problems. Recall that the goal of this project is to measure differences in compliance costs as they affect investment. Now imagine that all countries have exactly the same environmental regulations and associated compliance costs, and that all countries grandfather existing polluters. Furthermore, imagine that some countries receive more investment than others, perhaps even just by chance. In this situation, the countries receiving more investment will, as a consequence of the grandfathered regulations, have higher average compliance costs than the countries receiving less investment. An empirical test of the effect of compliance costs on investment will measure a spurious positive correlation: investment will seemingly flow to the high-cost countries, even though all states have the same regulations and there can be no true effect.

What is worse, now suppose that some countries have more stringent regulations for new investment than others. In this case *less* investment flows to those countries with stringent regulations, and more flows to the countries with lower costs for new investment. If all countries' regulations are nevertheless tighter for new investment than existing investment, the country receiving the new investment will have the higher measured costs. In this example, the apparent flow of investment to the high-cost countries will be even more dramatic, even though the true effect is that high-cost countries deter investment.

Because of the tendency for environmental compliance costs to vary by capital vintage – higher costs for newer capital – an international survey of average compliance costs by country will not be sufficient to estimate the effect of compliance costs on investment. Rather, we need a survey that includes questions about the age of manufacturing equipment.

Note, however, that this is not an easy concept. Manufacturing facilities are built over many years, updated and retrofitted periodically, and occasionally undergo drastic remodeling. In the U.S., where the laws explicitly grandfather existing sources of pollution, a great deal of controversy and litigation surrounds cases where manufacturers extensively retrofit existing sources of pollution and claim protection from the new more stringent standards.[2]

Problem #3: Industrial Composition

The third major conceptual hurdle facing surveys of compliance costs is that they vary tremendously by industrial composition. The average pollution abatement cost in a region, by production worker, or per dollar of value added, will be largely a function of the industrial composition of that region. Even if each region has identical regulations, but some regions have more pollution-intensive industrial compositions than others, the regions with more polluting manufacturers will have higher abatement costs. In this case, there will be no measurable effect of environmental regulations on investment.

Worse, suppose some regions have low compliance costs and *attract* polluting manufacturers. In this case, regions with lax regulations will have more pollution-intensive industrial compositions, and may well have *higher* average compliance costs than regions with more stringent standards. Here the measured effect of environmental regulations and investment will be spuriously positively correlated.

By far the closest approximation to the ideal data comes from the United States Census Bureau's Pollution Abatement Costs and Expenditures (PACE) survey, collected from 1972 to 1994, to be resumed in 2000, and from Statistics Canada's Environmental Protection Expenditures survey, begun in 1995. In the next section we describe an approach to estimating compliance cost differences across U.S. states using the PACE data, and use those estimates to examine the effect of costs on international investment.

An Example Using U.S. Data

Many researchers have relied on the PACE survey to construct indices of environmental costs in U.S. states.[3] The survey collected data from manufacturing establishments about their pollution abatement operating and capital costs. Most commonly, studies use these costs divided by some measure of state economic activity, such as total employment or gross state product.[4] The most significant problem with such measures is that they fail to adjust for differences in states' industrial compositions. States that have pollution-intensive industries will incur high pollution abatement costs, whether or not they have stringent regulations. Ideally, one would use the pollution abatement costs in the relevant industry as an index of regulatory stringency.

One method might be to use the plant-level data to estimate how much manufacturers pay for pollution abatement in each state, holding constant characteristics of the manufacturer, including its industry. Levinson (1996) contains such an estimate. It uses the raw, establishment-level PACE data from 1988, merged with data on plant characteristics from the 1987 Census of Manufactures. (Unfortunately the PACE data were not collected in 1987.) It then regresses the log of gross pollution abatement operating costs on the log of the book value of capital, the log of the number of production workers, the log of value added, a dummy for new plants, dummies for 4-digit SIC codes, and individual state dummies:

$$\ln(PACE) = \beta_0 + \beta_1 \ln(K) + \beta_2 \ln(L) + \beta_3 \ln(value\ \ added) + \beta_4 \ln(age) + \beta_5 \ln(4\text{ - }digit\ \ industry) + \beta_6 \ln(state) + \varepsilon \quad \textbf{(1)}$$

Where K is the book value of capital, and L is the number of production workers at the plant.

Note that this specification can be interpreted as an inverted Cobb-Douglas production function, where output (value added) is a function of capital (*K*), labor (*L*), and pollution (*P*), and dummy variables for new plants, industries, and states $Y = A \bullet K^{\beta 1} \bullet L^{\beta 2} \bullet P^{\beta 3}$. The model in equation **(1)** substitutes pollution abatement, which is measured by the PACE data, for pollution, which is unobservable, takes the logarithm of both sides, and inverts the function to estimate abatement as a function of the other variables.

The results are reproduced in Table 8.1, and the coefficients on the state dummy variables are interpreted as measures of compliance costs. A large state coefficient indicates that plants in that state spend more on pollution abatement operating costs, all else equal. Because the omitted state, New York, has relatively high costs, all of the statistically significant coefficients on other states are negative.[5] In many ways, the coefficients in Table 8.1 represent *exactly* what we would like to know in order to examine the effect of environmental costs on trade: They tell us how compliance costs for similar manufacturers differ across jurisdictions.

There are, however, several remaining problems with this measure of compliance costs. First, respondents to the PACE survey presumably have a great deal of difficulty reporting the true economic costs of pollution abatement, including efficiency gains or losses due to input substitution or altered production processes. At best, they provide direct dollar amounts spent on pollution abatement, less some amount for "cost offsets" due to efficiency gains, reported in recent PACE surveys. As a consequence, the plant-specific abatement operating costs may overstate or understate true compliance costs, and the dummy variable coefficients in Table 8.1 may contain an associated bias. If the respondents' error estimating their compliance costs is distributed randomly, it will amount to little more than standard measurement error, and will bias associated cost coefficients downwards unless appropriately treated, perhaps via instrumental variables. If, however, the error is systematic in some way, it poses a more complicated problem.

Second, the coefficient on the state dummy variable measures how much more a plant would have to spend on pollution abatement if it located in that state rather than in the omitted state, holding constant capital, labor, value added, and industry. But it is unlikely that plants locating in two different

TABLE 8.1. Industry-Specific Environmental Compliance Costs in the United States

Dependent Variable: In (gross pollution abatement operating costs)

Variable	Coefficient	Std. Error	Variable	Coefficient	Std. Error
In(Capital)	0.545*	0.016	MS	-0.255*	0.123
In(Production Workers)	0.439*	0.022	MT	0.110	0.273
In(Value Added)	0.084*	0.016	NC	-0.144	0.080
New Plant Dummy	0.060	0.061	ND	-0.566	0.384
AL	-0.035	0.094	NE	-0.196	0.144
AR	-0.072	0.103	NH	-0.276	0.178
AZ	-0.232	0.155	NJ	0.117	0.077
CA	-0.150*	0.064	NM	-0.500	0.322
CO	-0.384*	0.140	NV	-0.239	0.348
CT	-0.001	0.097	NY	na	na
DE	0.273	0.194	OH	0.056	0.067
FL	0.022	0.095	OK	-0.396*	0.120
GA	-0.194*	0.084	OR	0.122	0.110
IA	-0.034	0.104	PA	0.022	0.067
ID	-0.004	0.190	RI	-0.247	0.148
IL	0.055	0.067	SC	-0.184	0.096
IN	0.013	0.078	SD	-0.020	0.264
KS	-0.330*	0.115	TN	-0.078	0.088
KY	0.065	0.101	TX	-0.151*	0.071
LA	-0.102	0.107	UT	-0.494*	0.177
MA	-0.109	0.086	VA	-0.097	0.093
MD	0.148	0.108	VT	-0.111	0.220
ME	-0.041	0.163	WA	-0.182	0.107
MI	0.084	0.076	WI	-0.186*	0.078
MN	-0.209*	0.091	WV	-0.115	0.143
MO	-0.195*	0.091	WY	-0.412	0.365

n=11034, d.f.=1-565 R-Squared=.74
*Statistically significant at 5%. Includes dummy variables for 4-digit SIC codes.
Reprinted from Levinson (1996).

states would hold all of those other factors constant. Given that manufacturers can respond to regulations in ways aside from spending more on pollution abatement, this measure may overstate true compliance cost differences.

Still, the coefficients in Table 8.1 come as close as we can imagine to our ideal measure of compliance cost differences across regions, by holding constant other observable characteristics of firms, including size, industry, and age. Unfortunately, however, it is impossible to estimate versions of Table 8.1 for many years, as the U.S. Census Bureau has not retained the historical, disaggregate, plant-level data. Thus it is not possible to use these data to construct a time series of *changes* in relative compliance costs. And consequently, it is impossible to control for unobserved characteristics of regions that may be correlated with both compliance costs and investment, as outlined in the section above labeled "*Problem #1*." Using this single cross-section of compliance cost coefficients, Levinson (1996) concluded that there does not seem to be evidence that states with high relative compliance costs experience fewer new plant births, but could not rule out biases due to this unobserved heterogeneity.

As an alternative, it is possible to use the aggregate published PACE data to generate a long panel of compliance cost data. Although the U.S. Census Bureau does publish annual abatement costs by state and industry, many of the observations are censored to prevent disclosure of confidential plant-level information. Instead, Levinson (1999) proposes an alternative index.

The alternative index compares the average abatement costs in each state, across all industries, to a *prediction* of abatement costs by state, based on national average abatement costs by industry, and on and each state's industrial composition.[6]

Denote the average abatement costs per dollar of output as

$$S_{st} = \frac{P_{st}}{Y_{st}} \tag{2}$$

where P_{st} is pollution abatement operating costs in state s in year t, and Y_{st} is the total manufacturing sector's value added in state s in year t. This calculation, S_{st} is the type of aggregate measure of compliance costs commonly used. It fails to adjust for the industrial composition of each state, and so it probably exaggerates compliance costs in states with relatively more polluting industries and understates costs in clean states.

Now compare **(2)** to a *prediction* of pollution abatement costs:

$$\hat{S}_{st} = \frac{1}{Y_{st}} \sum_{i=20}^{39} \frac{Y_{ist} P_{it}}{Y_{it}} , \tag{3}$$

where i indexes 2-digit SIC codes for manufacturing industries, 20 through 39,[7] Y_{ist} is industry i's value added in state s at time t, Y_{it} is the national value added of industry i, and P_{it} is national pollution abatement operating costs for industry i. $_{st}$ is a weighted average of pollution abatement operating costs (per dollar of value added), in which the weights are relative shares of output for each industry in state s at time t.

The alternative, industry-adjusted measure of compliance costs in Levinson (1999) is the ratio of average state expenditures in **(2)** to the predicted state expenditures in **(3)**

$$S^*_{st} = \frac{S_{st}}{\hat{S}_{st}} . \tag{4}$$

When S_{st}* exceeds 1, industries in state s at time t have spent more on abatement than the same industries spent in other states. When S_{st}* falls below 1, industries in state s at time t spent relatively less on abatement. One interpretation of S_{st}* is that states with high S_{st}* have more stringent regulations than states with small S_{st}*. [8]

Keller and Levinson (1999) use this index, S_{st}*, to try to assess the effects of compliance costs on foreign direct investment. As a measure of international investment flows, they use the value of gross property, plant and equipment belonging to foreign manufacturers from the U.S. Department of Commerce, Bureau of Economic Analysis (BEA), *Foreign Direct Investment in the United States*.

Table 8.2 contains an example of this type of analysis. The first column presents the average industry-adjusted index S*, from 1977 to 1994, for various combinations of states. The average S* for the five states with the lowest compliance costs is 0.64, indicating that those states have 64 percent of the compliance costs they could be expected to have given their industrial compositions. The average S* for the highest five states is 1.59, suggesting their compliance costs are 59 percent higher than would be predicted by their industrial compositions. Rows (3) and (4) of Table 8.2 present that same information for the 20 lowest and highest states, respectively.

Column (2) of Table 8.2 presents the average value of gross property, plant and equipment (PP&E) investment by foreign-owned manufacturers from 1977 to 1994, for all manufacturers, scaled by each states Gross State Product (GSP — the state equivalent of Gross Domestic Product). On average, in the five states with the lowest compliance cost indices, FDI is a *lower* fraction of GSP than in the five states with the highest indices. The same is true for the bottom panel of Table 8.2, which compares the states with the 20 lowest and 20 highest indices.

Because most pollution abatement expenses are incurred by a few industries, the results in column (2) may be dampened by the vast majority of industries that have relatively minor abatement expenditures. Consequently, column (3) examines SIC 28, chemicals and allied products, as an example of a pollution-intensive industry. Of polluting industries, this industry has the most consistently reported uncensored FDI data in the BEA publications. However, as column (3) shows, for the chemical industry the counter-intuitive gap between the most and least costly states is even *larger*. It appears as if investment flows to states with high environmental compliance costs, and that pollution intensive investment has an even greater tendency to migrate to high cost states.

This is the type of result that pervades the empirical literature on trade and environmental regulations. It suggests no evidence of a pollution haven effect, no industry relocation response to environmental regulations, and no need to worry about a race to the bottom in environmental standards. In fact, it suggests that pollution-intensive industries are *attracted* to states with high compliance costs. We fear, however, that results are due to the endogeneity of pollution regulations and compliance costs. States that attract relatively little polluting manufacturing need not enact stringent regulations, whereas states that *do* attract polluting manufacturing may respond by enacting more

TABLE 8.2. Summary Statistics — Averages 1977-1994

	Industry-adjusted index of abatement costs [S*]	Foreign direct investment (PP&E as a fraction of Gross State Product)	
		All manufacturing	SIC 28 -- chemicals
	(1)	(2)	(3)
Avg. for lowest 5	0.64	0.0233	0.0073
Avg. for highest 5	1.59	0.0604	0.0335
Avg. for lowest 20	0.75	0.0273	0.0112
Avg. for highest 20	1.33	0.0487	0.0257

Source: Author's calculations.

costly regulations. Consequently, comparing investment in costly and inexpensive states, we find that more investment flows to the costly states. Furthermore, when we look at polluting industries in particular, the endogeneity bias becomes worse.

This bias is an example of what above we described as "*problem #1*." The clearest solution comes from comparing compliance costs over time. In other words, rather than examining *levels* of investment and compliance costs, we look at *changes* in investment and compliance costs.

Table 8.3 thus examines *changes* in compliance costs and FDI. We calculated the average stringency and FDI for the first five years of the data (1977-1981) and for the last five years (1990-1994), and subtracted the latter from the former. The five states whose average S^* declined most over the 18 year period had an average change of -0.597. For these states, average annual value of PP&E foreign direct investment as a fraction of GSP grew by 4.09 percentage points. In contrast, the five states whose S^* increased most over the period saw their FDI as a fraction of GSP grow by 2.83 percentage points. Taken at face value, this comparison suggests that states that became more costly attracted less FDI.

Column (3) of Table 8.3 examines the chemicals industry, SIC 28. Here, the distinction between the five states whose S^* increased most and the five whose S^* declined most is more stark than in column (2). Furthermore, these distinctions are robust to the set of states examined (20 greatest and smallest index changes, 5 greatest and smallest, or other combinations in between.)

Although Table 8.3 appears to present strong evidence of a deterrent effect of environmental regulations, we do not want put too much emphasis on that result. The analysis in Table 8.3 controls for no other observable characteristics of states that may be correlated with both compliance costs and foreign investment.[9] Rather, we think the most important lesson comes from comparing Tables 8.2 and 8.3. Table 8.2 finds a positive correlation between *levels* of investment and *levels* of costs, while Table 8.3 demonstrates a negative correlation between *changes* in investment and *changes* in costs. From the comparison, one can see quite clearly the bias associated with cross-sectional analyses of investment and compliance costs. The lesson, we think, is clear: Any study that relies on a measure of stringency or compliance costs across regions at a single point in time is likely to suffer from large and unavoidable endogeneity biases.

To conduct a study like this at the international level is at this time, unfortunately, impossible. No agency that we know has collected compliance costs by industry internationally, let alone over time. Therefore, to examine the effect of compliance cost differences across *countries* rather than *states*, one is forced to use more indirect measures of costs. In the next section, we discuss some of the existing international measures of standard stringency, with an eye towards the lessons learned from studying U.S. states.

International measures of standard stringency

Analyses of international environmental standards all face a similar problem: they lack data. This data shortage impedes efforts to explore the degree, causes, and effects of international differences in environmental standards. And, this information shortage seriously handicaps efforts to analyze the links between international trade and environmental protection. Lacking compliance cost data, discussions of environmental standards often seem to resort to anecdotes. At hearings on the proposed "International Pollution Deterrence Act of 1991," U.S. Senator Max Baucus (D-MT) described pictures of "poorly buried drums of hazardous waste" in Mexico as evidence of "different levels of environmental protection around the world" that gave a "competitive advantage" to nations that maintain "inadequate environmental protection" (U.S. Congress, Senate 1991). Senator Earnest Hollings (D-SC), at 1994 hearings on trade and the environment, described "mounds of lead and...shanty homes" as evidence of "weak standards" (U.S. Congress, Senate 1994). This section reviews past efforts to construct systematic data on international standard stringency, and drawing on the lessons learned from earlier analyses, proposes general guidelines for assembling such data.

The United States Government

The "International Pollution Deterrence Act of 1991," sponsored by U.S. Senator David Boren (D-OK), included provisions for a comprehensive assessment of international environmental standards. The proposed legislation, though never enacted, called for:

> "... a pollution control index for each of the top fifty countries identified by the Office of Trade and Investment of the Department of Commerce based on the value of exports to the United States from that country's attainment of pollution control standards in the areas of air, water, hazardous waste and solid waste as compared to the United States. The purpose of this index is to measure the level of compliance within each country with standards comparable to or greater than those in the United States. The Administrator shall analyze, in particular, the level of technology employed and actual costs incurred for pollution control in the major export sectors of each country formulating the index."

According to U.S. Senator Max Baucus (D-MT), a major goal of the legislation was to "level the playing field" with nations that have "inadequate environmental protection."

Whatever one thinks of its motives, the proposed index could have provided useful data concerning international differences in environmental standards and pollution-control costs. And, to be consistent with its goals, the law would have had to require that the EPA estimate these cost differences at regular intervals, generating exactly the type of time-series data needed. Despite its limitations and eventual failure to gain the required support, this proposal stands out as a particularly ambitious (perhaps overly ambitious) proposal for measuring international differences in environmental standards.

The United Nations

The United Nations has sponsored two efforts to document global differences in environmental standards. The first emerged in connection with the 1972 Stockholm Conference — known formally as the United Nations Conference on the Human Environment. Conference leaders commissioned the United Nations Conference on Trade and Development (UNCTAD) to study the ways in which differences in environmental standards could influence trade patterns. In investigating the differences in environmental policies, UNCTAD researchers sent questionnaires to 145 member states and received a total of 40 replies (Walter and Ugelow, 1979). Ten of the responses provided small amounts of useful data. Fourteen offered helpful but not outstanding information. Another 14 supplied detailed descriptions. (Two of the responses apparently eluded classification.) The analysts ranked the responses along a continuum from "tolerant policies" (i.e., very little control over polluting activity) to "strict policies" (i.e., extensive and articulated controls over polluting activity). The ranking used the United States as a benchmark for "strict."

While the UNCTAD survey deserves credit for producing useful data, it also leaves room for improvement. First, the researchers experienced a very low response rate — 28 percent responded, 10 percent responded in detail. Second, the survey, evidently in an effort not to overburden the respondents, included a very small number of questions.[10] Third, UNCTAD evidently asked open-ended questions that allowed strategic responses — so that countries could rightly or wrongly portray themselves as environmentally active. Fourth, analysts' judgements played a large role in assembling the scale. Fifth, the questionnaire's ability to assess enforcement seems particularly limited. Sixth, the results apparently did not help researchers to estimate specific differences in environmental costs. Despite its flaws, UNCTAD's survey marked an important first attempt to characterize international differences in environmental standards.

Another U.N.-sponsored effort to assess environmental standards occurred in association with the 1992 "Rio Earth Summit." In preparation for that summit, the United Nations Conference on Environment and Development (UNCED) asked participants to describe their environmental policies. The reports proved noteworthy for many reasons. First, they formed a thorough compendium of environmental policies.[11] Second, the descriptions of "Past and Present Capacity Building Initiatives" depicted great progress in nations' that many presumed had not formulated meaningful policies (e.g., India, Burkina Faso, the Russian Federation and Venezuela). Third, only 15 of 169 reports met the original deadline. Only 57 met the extended deadline. The response rate might suggest some countries lacked either the will or the ability to provide the reports. Fourth, political considerations seem to have influenced some of the reporting.[12] Fifth, although the committee requested data on "enforcement procedures," national reports contained very little information on the extent to which they enforce the current laws. Many countries (e.g., Suriname, Nepal and Ethiopia), in other sections of their reports, alluded to a need for stricter enforcement. Finally, like the UNCTAD study that preceded it, the UNCED analyses did not allow for a calculation of regulation-based differences in production costs.

TABLE 8.3. Changes in Average Pollution Abatement Costs and FDI 1977-1981 to 1990-1994

	Change in Industry-adjusted index on abatement costs [S*] (1)	Change in foreign direct investment (PP&E as a fraction of Gross State Product) All manufacturing (2)	SIC 28 - chemicals (3)
Largest 5 declines	-0.597	0.0409	0.0428
Largest 5 increases	0.447	0.0283	0.0004
Largest 20 declines	-0.230	0.0394	0.0217
Largest 20 increases	0.190	0.0255	0.0029

Source: Author's calculations.

The Inter-American Development Bank

Writing for the Inter-American Development Bank (IADB), Raul Branes (1991) authored a detailed investigation of its member countries' environmental policies. Its stated purpose was to "provide a comprehensive picture of the way in which environmental management is organized in (member) countries... (and to provide) information on legal and institutional aspects of environment issues..." for the public and for IADB employees. The report evaluated eight types of environmental legislation for ten nations.[13] Like the U.N. projects and the proposed U.S. project, the Branes study stands out as a commendable first step in a comprehensive assessment of international environmental standards. Like those studies, this one did not provide specific information about regulation-based differences in production costs. Neither did it evolve into a steady flow of environmental intelligence. Branes himself also identified a number of caveats that should accompany any interpretation of his findings. First, he warns that the laws differ dramatically from one country to another. Some countries' environmental legislation deal only in superficialities, while others' laws address deep roots. Second, he cautions that within most nations, the environmental laws lack coherence. Third, Branes points out that "general enforcement of environmental legislation in the countries of the region has run into serious problems of efficiency and efficacy." He defines "efficacy" as the extent to which the law is observed and "efficiency" as the extent to which the law is suitable for meeting the identified needs. Widespread inefficacy, he claims, results mostly from "shortages of human, material and financial resources" and from shortcomings in judicial enforcement."

Other Efforts

In addition to the efforts described above, a number of other organizations have tried to construct systematic evaluations of international environmental standards and laws.

UNEP/WHO – In 1992, The World Health Organization (WHO), in conjunction with the United Nations Environment Programme (UNEP), published *Urban Air Pollution in the Megacities of the World,* a review of the air quality in 25 of the world's largest cities. The data, taken from the Global Environmental Monitoring System (GEMS), reported on each city's ambient concentrations of sulphur dioxide, suspended particulate matter, oxides of nitrogen, lead, ozone, and carbon monoxide. The report, in addition to evaluating the overall air pollution situation and control measures, provides specific recommendations for each city.

NRDC and CAPE 21 – In 1994, The Natural Resources Defense Council (NRDC) and the Campaign for Action to Protect the Earth (CAPE 21), released *Four in '94: Two years after Rio – Assessing National Actions to Implement Agenda 21.* The report evaluated the extent to which 73 nations, all participants in the 1992 Rio Earth Summit, had carried through on four agreements made at the Rio conference: reducing lead exposures, cleaning up freshwater resources, ratifying and implementing the Basel convention, and protecting natural rivers. The study identified a variety of successes and failures. Prior to releasing *Four in '94,* the two groups had published *One Year After Rio,* which broadly examined what nations had done to follow up on the environmental summit.

OECD – The Organization for Economic Co-operation and Development (OECD) has, since 1985, published The Compendium of Environmental Data on a bi-annual basis. Most of the information comes from questionnaires developed by the OECD Group on the State of the Environment, that are sent to member countries. The 1999 Compendium provides somewhat detailed reports on the quality of air, land, water, forests, wildlife waste management and risks. It also surveys critical trends in the energy, transport, industry and agriculture sectors, and key developments in environmental management.

Smaller scale attempts to describe international differences in environmental standards frequently appear in periodicals such as the International Environment Reporter, Environmental Policy and Law, and the International Journal of Environmental Affairs.

Lessons Learned

Perhaps obviously, it is difficult to summarize all previous attempts to compare countries' environmental standards. However, we feel it is fair to make the following observations. First, no survey has been repeated. This impedes efforts to control for unobserved attributes of countries by comparing changes in regulations and changes in trade or investment. Second, many of the surveys asked numerous questions unrelated to manufacturers' costs. These questions likely reduced response rates and distracted respondents from the central issue. Consequently, if we can make a few suggestions for future attempts to measure international *compliance cost* differences, they would be the following.

Repetition – First, because repeated observations over time are so valuable for these types of analyses, plan on (a) repeating the survey periodically in the future, and (b) repeating, as closely as possible, a few of the questions asked by previous studies. The repetitions allow for comparisons across time and would help to reveal influences that might otherwise go unobserved.

Focus – Second, focus on questions related to manufacturers' compliance costs. If the survey is to have other goals besides assessing the trade-environment tradeoff, then other questions may make sense. But if the sole goal is to measure regulatory effects on trade, then it makes no sense to ask countries or companies about their recycling programs, national parks, endangered species protection programs, or automobile emissions standards. The meager response rates to the UNCTAD and UNCED surveys suggest that many respondents lack the ability or the will to provide the desired data. Assessment efforts will certainly achieve better results if they focus their inquiry so as to simplify the response process and minimize the demands imposed on respondents. A limited number of questions on a limited number of topics will work best. The survey could certainly be designed to provide more data than the original UNCTAD study. Still, it should recognize the limited results achieved by ambitious surveys and the success of focused efforts such as NRDC's *Four in '94,* which set out to answer four questions for many countries.

Caution – Third, if the survey is to consider issues other than manufacturers' compliance costs, it should be expanded with caution. Broad analysis of regulation, if it is to be useful, involves extensive effort. For example, a thorough survey of environmental legislation would not only identify the laws that are on the books, it would also consider: institutional arrangements, including the role of regulatory agencies, the influence of NGOs and the possibility of environmentally based law suits; the distribution of laws, so that one would know whether the regulations apply to the entire nation or only to some region; and, enforcement and compliance, a topic which has provided the greatest challenge to earlier surveys.[14] Attempts to expand a survey's focus, beyond a measurement of what it costs firms to comply with environmental regulations, should be considered with caution as they necessarily involve a great deal more research – and a lower response rate – than one might originally suspect.

Concluding Remarks

In the end, two major lessons stand out. The first is the critical importance of measuring standard stringency in more than one year, in order to study the effect of changes in standard stringency on changes in international trade. Cross-section analyses, done at one point in time, are doomed by omitted variable bias and the endogeneity of the standards. While in practice this means it will be necessary to collect data for many years, in the short run it could involve replicating international surveys conducted earlier, or at least the key successful parts of such surveys. The second critical lesson is that comparisons of national environmental policies can easily be distracted by lots of issues unrelated to manufacturer costs.

If the goal is studying the effect of policies on trade, only those policies directly related to manufacturers' costs need to be included.

NOTES

1. *New York Times*, October 8, 1980, cited in Leonard (1988).

2. "EPA Finds Broken Rules At Coal Plants, Officials Say" *New York Times*, July 16, 1999, Matthew L. Wald.

3. The entire PACE survey, and summary statistics for 1994, can be seen at ***http://www.census.gov/prod/2/manmin/ma200x94.pdf***.

4. Crandall (1993), Friedman *et al.* (1992), or Co and List (1998). Consulting firms specializing in industrial siting decisions have also relied on such simple indices of environmental regulatory stringency (Alexander Grant & Co., 1985).

5. The asterisks in Table 8.1 merely indicate where the relevant coefficients are statistically different from zero. More importantly, however, many coefficients are statistically different from *each other*.

6. The index uses pollution abatement operating expenses rather than capital expenses, though both are available from PACE data. Operating expenses may be easier for survey respondents to identify separately, while capital expenses are likely to be more difficult to segregate from investment that is unrelated to pollution abatement. Also, abatement capital expenditures peak when new investment occurs, which means that states with relatively more investment have relatively higher abatement capital costs, regardless of the stringency of those states' environmental laws.

7. The PACE survey omits SIC code 23 (apparel) because it is relatively pollution-free.

8. For more details about the index, and summary statistics for individual states over time, see Levinson (1999). Interested readers can also find the data for this index at ***www.georgetown.edu/faculty/aml6/index2.htm***.

9. In Keller and Levinson (1999), when other observable characteristics of states are accounted for, the measured deterrent effect of environmental costs is mitigated but not completely erased.

10. The original questionnaire has proved impossible to locate. Later comment on the questionnaire suggests that the original contained only ten questions (Blond, 1974). A 1979 follow up survey seems to have included only 3 questions (UNCTAD, 1979).

11. Paragraph 16-1-a of the "Guidelines for National Reports" provided the following instructions for report-writers:

> "Outline the major approaches that have been developed and implemented at the national level over the past two decades to address environmental and resource problems as they relate to development, and indicate the principles that guide policy, the key goals, targets and priorities of development, and the policies, legislation and institutions that enable their achievement." (UNCED, 1992)

Paragraph 16-2-b instructed those reporters to describe "new departments or councils on the environment, establishment of rules and regulations and enforcement procedures" (UNCED, 1992).

12. For example, Haiti's report, written by allies of then-exiled President Aristide, stands out as a notable exception to the upbeat tone. Haiti reported that "a large number of institutions were set up to take care of the environment ... (but) the tasks are at times neglected... Their establishment was a result of clan struggles for control or embezzlement of capital. The functioning of these state bodies is often quite arbitrary."

13. The countries were Argentina, Bolivia, Brazil, Colombia, Chile, Ecuador, Paraguay, Peru, Uruguay and Venezuela. The environmental sectors were water, forests, wildlife, soils, marine and coastal ecosystems, non-renewable natural resources, human settlements and environmental sanitation. (Of course, not all of these can be expected to affect manufacturers' costs.)

14. One alternative to measuring enforcement and compliance is to measure performance. Hilton and Levinson (1998) have taken this approach by measuring lead levels in automotive gasoline. Unlike studies that examine rules concerning lead phase out, they measure the actual lead content of gasoline. The focus on actual performance eliminates the need for entry into the quagmire of measuring enforcement.

REFERENCES

Alexander, Grant & Co. various years. *Annual Study of General Manufacturing Climates of the Forty-eight Contiguous States of America.* Chicago: Alexander, Grant & Co.

Berman, E. and L. Bui. 1999. Environmental Regulation and Productivity: Evidence from Oil Refineries. mimeo. Boston, MA: Boston University.

Birdsall, N. and D. Wheeler. 1992. Trade Policy and Industrial Pollution in Latin America: Where are the Pollution Havens? *Journal of Environment and Development.* Winter 1993. 2: 1.

Branes, R. 1991. *Institutional and Legal Aspects of the Environment in Latin America, Including the Participation of Nongovernmental Organizations in Environmental Management.* Washington, D.C.: Inter-American Development Bank.

Co, C. and J. List. 2000. The Effects of Environmental Regulations on Foreign Direct Investment. *Journal of Environmental Economics and Management* 40(1).

Copeland, B.R. 1990. Strategic Interaction Among Nations: Negotiable and Non-negotiable Trade Barriers. *Canadian Journal of Economics*, 23: 64-108.

Crandall, R.W. 1993. *Manufacturing on the Move.* Washington, D.C.: The Brookings Institution.

Ederington, J. 1998. International Trade and Domestic Policies. mimeo. University of Miami.

Friedman, J., D.A. Gerlowski and J. Silberman. 1992. What Attracts Foreign Multinational Corporations? Evidence From Branch Plant Location in the United States. *Journal of Regional Science* 32(4): 403-418.

Greenstone, M. 1998. The Impacts of Environmental Regulations on Industrial Activity: Evidence from the 1970 and 1977 Clean Air Act Amendments and the Census of Manufactures. mimeo. University of California at Berkeley.

Hall, B. and M.L. Kerr. 1991. *Green Index: A State-by-State Guide to the Nation's Environmental Health*. Washington, D.C.: Island Press.

Hilton, F.G. and A. Levinson. 1998. Factoring the Environmental Kuznets Curve: Evidence from Automotive Lead Emissions. *Journal of Environmental Economics and Management*, 35 126-141.

International Trade Administration (ITA). various years. *Foreign Direct Investment in the United States*. Washington, D.C: United States Department of Commerce. Office of Trade and Economic Analysis.

Jaffe, A.B., et al. 1995. Environmental Regulations and the Competitiveness of U.S. Manufacturing: What does the Evidence Tell Us? *Journal of Economic Literature* 33(1): 132-63.

Kahn. M. 2000. United States Pollution Intensive Trade Trends from 1972 to 1992. mimeo. New York, NY: Columbia University.

Keller, W. and A. Levinson. 1999. Environmental Compliance Costs and Foreign Direct Investment Inflows to U.S. States. NBER (National Bureau of Economic Research) Working Paper #7369. Cambridge, MA: NBER.

Kolstad, C., and Y. Xing. 1997. Do Lax Environmental Regulations Attract Foreign Investment? University of California Santa Barbara Working Paper #6-95R February.

Leonard, H.J. 1988. *Pollution and the Struggle for the World Product*, Cambridge MA: Cambridge University Press.

Levinson, A. 1996. Environmental Regulations and Manufacturers' Location Choices: Evidence from the Census of Manufactures. *Journal of Public Economics* Vol. 62.

———. 1999. An Industry-Adjusted Index of State Environmental Compliance Costs. Cambridge, MA: NBER.

List, J.A. 1999. U.S. County-Level Determinants of Inbound FDI: Evidence from a Two-Step Modified Count Data Model. *International Journal of Industrial Organization*, forthcoming.

Low, P., ed. 1992. *International Trade and the Environment*. Washington, D.C.: The World Bank.

Low, P. and A. Yeats. 1992. Do 'Dirty' Industries Migrate? in Patrick Low, ed., *International Trade and the Environment*. Washington, DC: The World Bank.

Meagher, J.P. 1990. Environmental Protection and Industries in Developing Countries: The Case of India since Bhopal. *Georgetown International Environmental Law Review*, 3: 1-54.

NRDC (Natural Resources Defense Council) and CAPE 21 (Campaign for Actions to Protect the Earth). 1994. *Four in '94: Two Years After Rio – Assessing National Actions to Implement Agenda 21*. Washington, D.C.: NRDC.

OECD (Organisation for Economic Co-operation and Development). 1999. OECD Environmental Data – Compendium 1999. Paris: OECD.

Rauscher, M. 1997. *International Trade, Factor Movements, and the Environment.* Oxford and New York: Oxford University Press, Clarendon Press.

Statistics Canada. Environmental Protection Expenditures in the Business Sector, 1995. Item #16F0006XIE.

UNCED (United Nations Conference on Environment and Development). 1992. Nations of the Earth Report. Geneva: UNCED.

UNCTAD (United Nations Conference on Trade and Development). 1979. Walter Ingo mimeo.

U.S. Department of Commerce, Bureau of the Census. various years. Current Industrial Reports: Pollution Abatement Costs and Expenditures, MA-200. Washington, D.C.: U.S. Department of Commerce.

Walter, I. 1973. The Pollution Content of American Trade. *Western Economic Journal*, 11: 61-70.

———. 1982a. International Economic Repercussions of Environmental Policy: An Economist's Perspective. In Seymour J. Rubin and James R. Graham, eds. *Environment and Trade.* Totowa, NJ: Allanheld, Osmun, & Co..

———. 1982b. Environmentally Induced Industrial Relocation to Developing Countries. In Seymour J. Rubin and James R. Graham, eds. *Environment and Trade.* Totowa, NJ: Allanheld, Osmun, & Co.

Walter, I. and J.L. Ugelow. 1979. Environmental Policies in Developing Countries. *Ambio*, 8: 102-109.

Wheeler, D. and P. Martin. 1992. Prices, Policies and the International diffusion of Clean Technologies. In *International Trade and the Environment,* ed. P. Low. Washington, DC: The World Bank.

WHO (World Health Organization) and UNEP (United Nations Environment Programme). 1992. *Urban Air Pollution in Megacities of the World.* Cambridge, MA: Blackwell.

Contributors

Kym Anderson
School of Economics and Centre for International Economic Studies
University of Adelaide, Adelaide, Australia
Centre for Economic Policy Research, London

Richard E. Baldwin
Graduate Institute of International Studies, Geneva, Switzerland

Abiodun S. Bankole
Department of Economics, University of Ibadan
Ibadan, Nigeria

Neil Gandal
Tel Aviv University, Tel Aviv, Israel
University of California at Berkeley, Berkeley, CA

Mattias Ganslandt
Research Institute for Industrial Economics, IUI, Stockholm, Sweden
University of Colorado at Boulder, Boulder, CO

Francis (Hank) G. Hilton, S.J.
Loyola College in Maryland
Baltimore, MD

Arik Levinson
Economics Department, Georgetown University
Washington, DC

James R. Markusen
Department of Economics
University of Colorado at Boulder, Boulder, CO
National Bureau of Economic Research, Boston, MA
Centre for Economic Policy Research , London

Keith E. Maskus
Department of Economics University of Colorado at Boulder
Boulder, CO

Chantal Nielsen
University of Copenhagen and
Danish Institute of Agricultural and Fisheries Economics (SJFI)
Copenhagen, Denmark

E. Olawale Ogunkola
Department of Economics, University of Ibadan
Ibadan, Nigeria

Tsunehiro Otsuki
Development Economics Research Group, The World Bank
Washington, DC

T. Ademola Oyejide
Department of Economics, University of Ibadan
Ibadan, Nigeria

John S. Wilson
Development Economics Research Group, The World Bank
Washington, DC

STUDIES IN INTERNATIONAL TRADE POLICY

Studies in International Trade Policy includes works dealing with the theory, empirical analysis, political, economic, legal relations, and evaluations of international trade policies and institutions.

General Editor: Robert M. Stern

John H. Jackson and Edwin Vermulst, Editors. *Antidumping Law and Practice: A Comparative Study*

John S. Odell and Thomas D. Willett, Editors. *International Trade Policies: Gains from Exchange between Economics and Political Science*

Stephen V. Marks and Keith E. Maskus, Editors. *The Economics and Politics of World Sugar Policies*

J. Michael Finger, Editor. *Antidumping: How It Works and Who Gets Hurt*

Horst Herberg and Ngo Van Long, Editors. *Trade, Welfare, and Economic Policies: Essays in Honor of Murray C. Kemp*

David Schwartzman. *The Japanese Television Cartel: A Study Based on* Matsushita v. Zenith

Alan V. Deardorff and Robert M. Stern, Editors. *Analytical Perspectives and Negotiating Issues in the Global Trading System*

Edwin Vermulst, Paul Waer, and Jacques Bourgeois, Editors. *Rules of Origin in International Trade: A Comparative Study*

Alan V. Deardorff and Robert M. Stern, Editors. *The Stolper-Samuelson Theorem: A Golden Jubilee*

Kent Albert Jones. *Export Restraint and the New Protectionism: The Political Economy of Discriminatory Trade Restrictions*

Alan V. Deardorff, James A. Levinsohn, and Robert M. Stern, Editors. *New Directions in Trade Theory*

Robert Baldwin, Tain-Jy Chen, and Douglas Nelson. *Political Economy of U.S.–Taiwan Trade*

Bernard M. Hoekman and Petros C. Mavroidis, Editors. *Law and Policy in Public Purchasing: The WTO Agreement on Government Procurement*

Danny M. Leipziger, Editor. *Lessons from East Asia*

Tamin Bayoumi. *Financial Integration and Real Activity*

Harry P. Bowen, Abraham Hollander, and Jean-Marie Viaene. *Applied International Trade Analysis*